MASTERING ESSENTIAL MATH SKILLS
Graph Paper Notebook

Great for All Geometry Classes

Also Includes a Handy
MATH RESOURCE CENTER
STARTING ON PAGE 117

Richard W. Fisher

Math
Essentials
LOS GATOS, CALIFORNIA

Mastering Essential Math Skills Graph Paper Notebook

Manufactured in the United States of America

ISBN: 978-1-7335018-7-3

1st printing 2019

Math Essentials
P.O. Box 92
Marina, CA 93933
Ph 408-314-4573
www.mathessentials.net
math.essentials@verizon.net

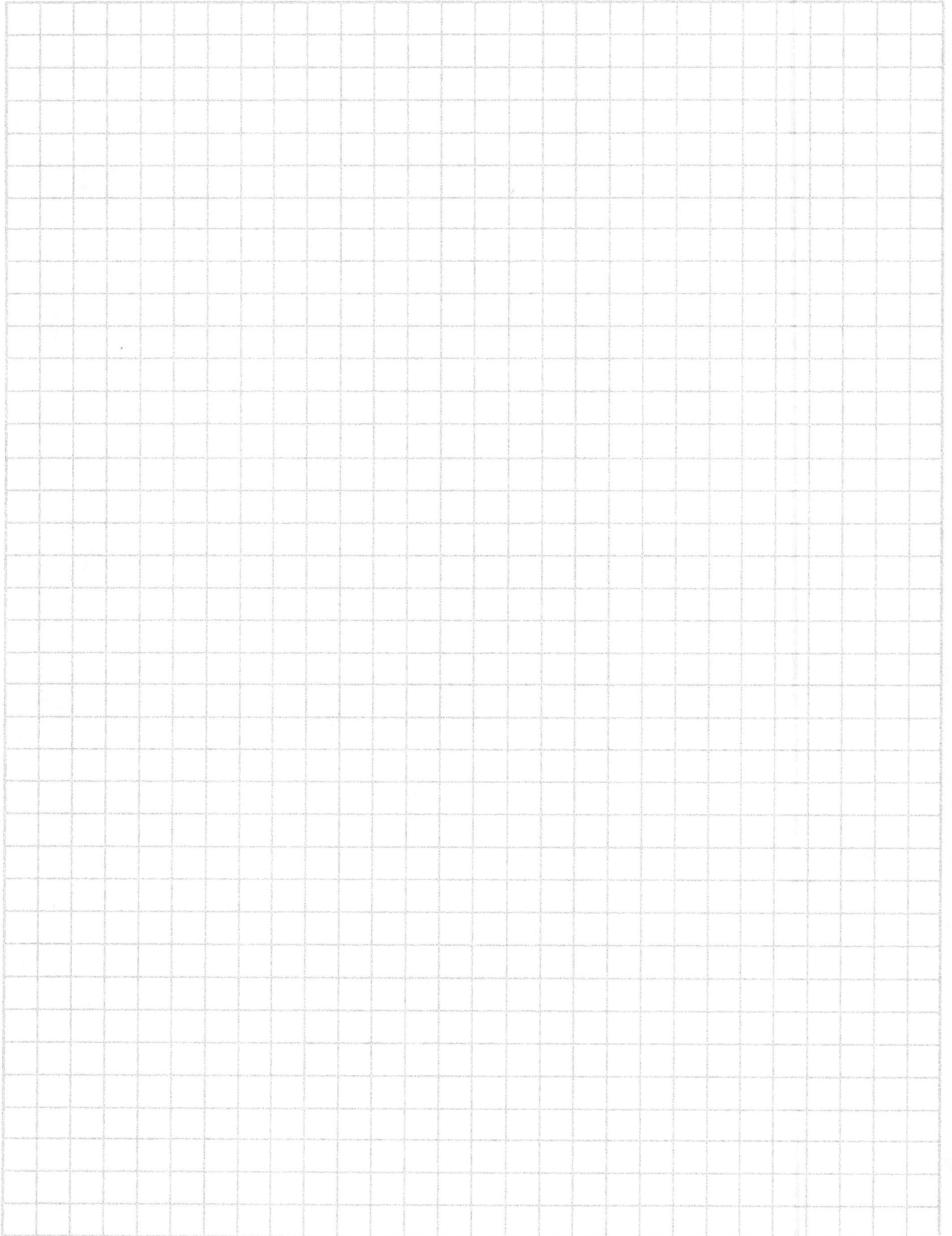

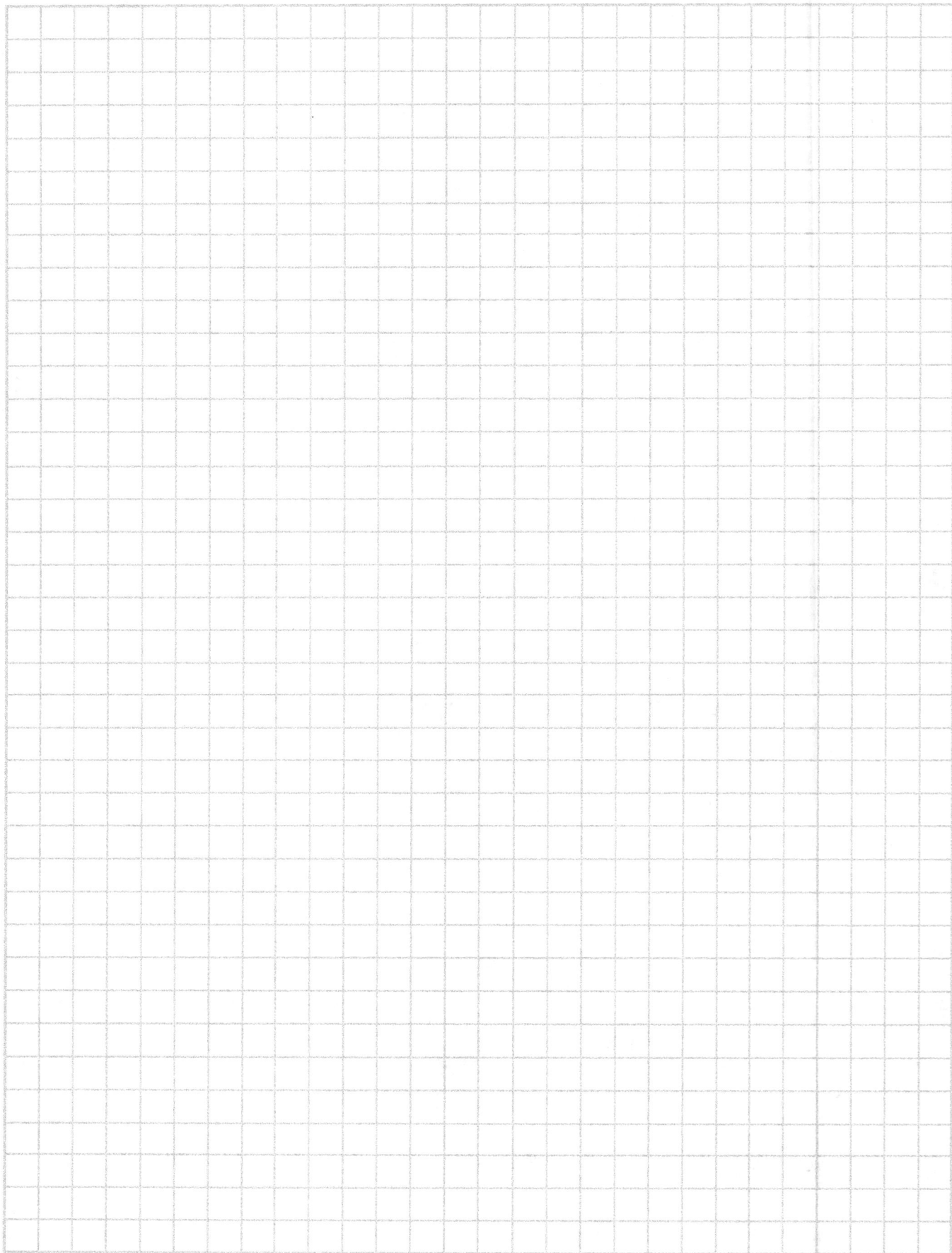

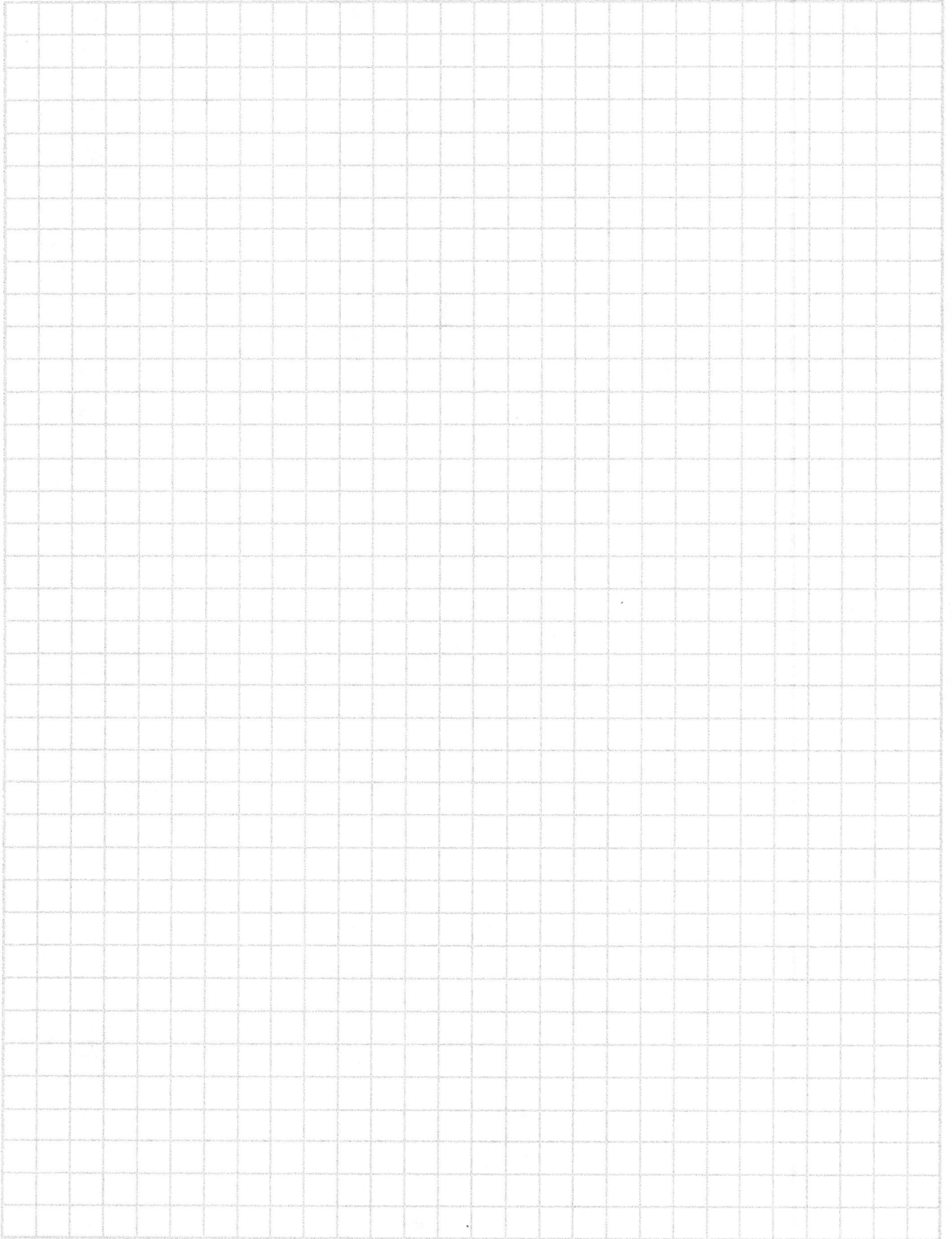

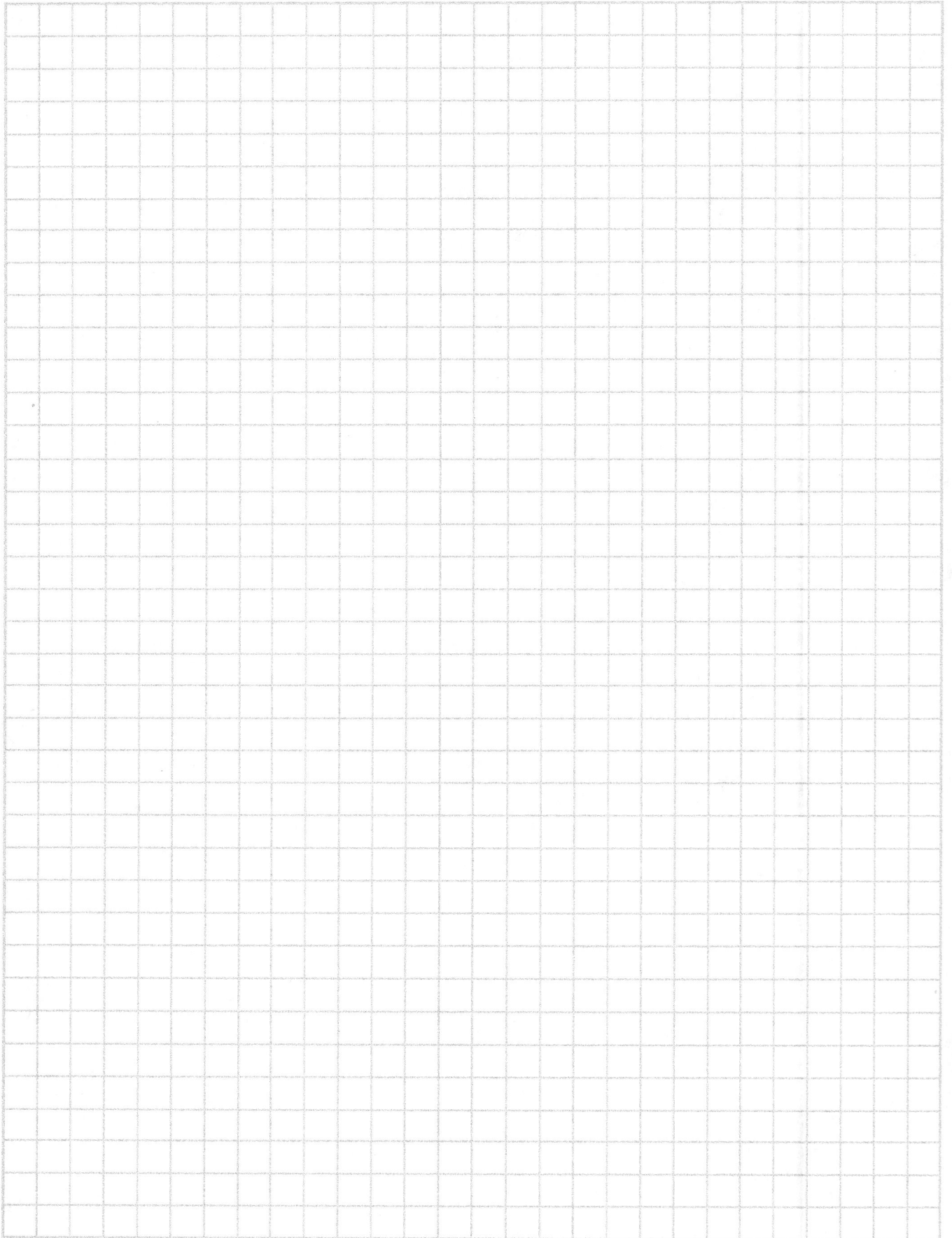

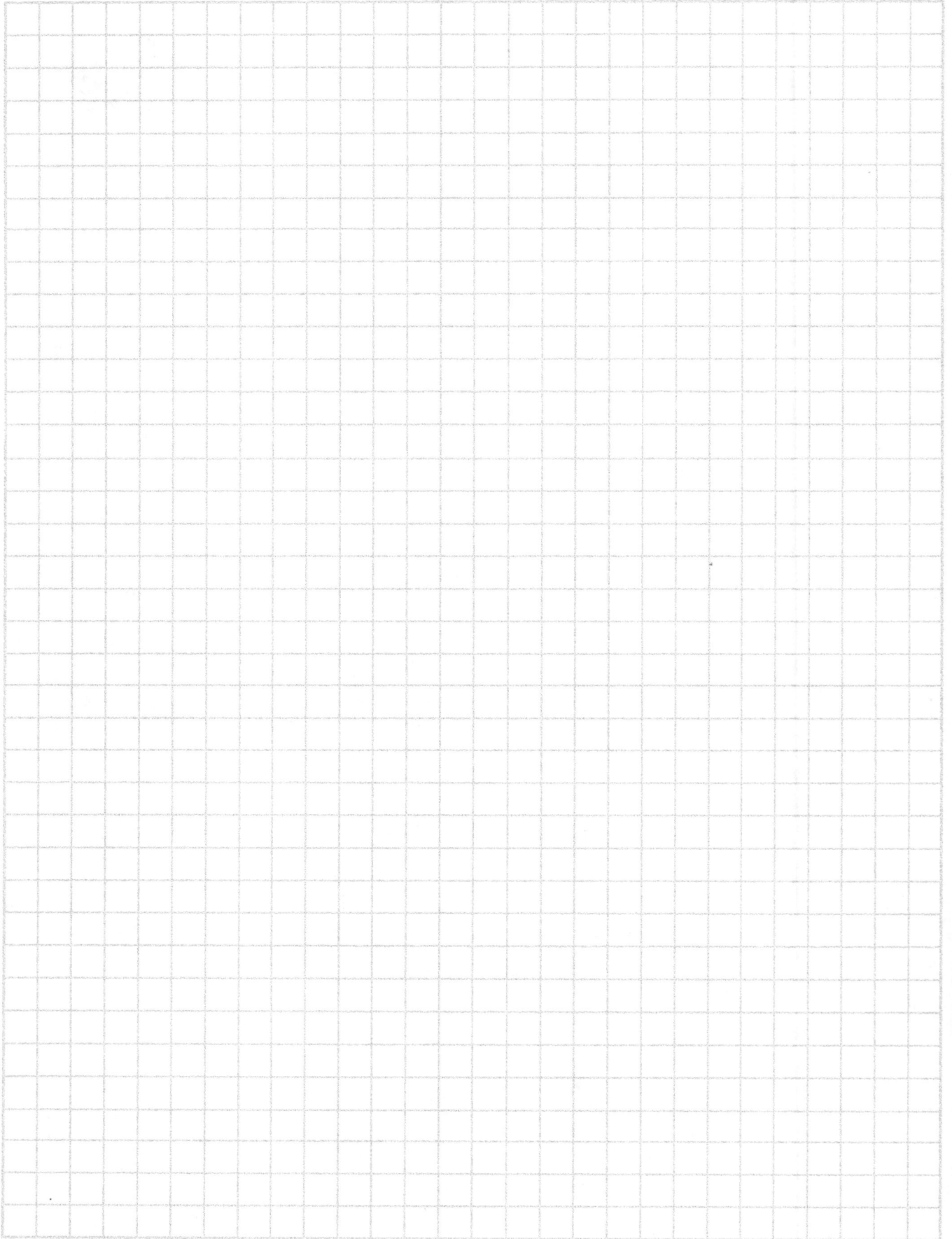

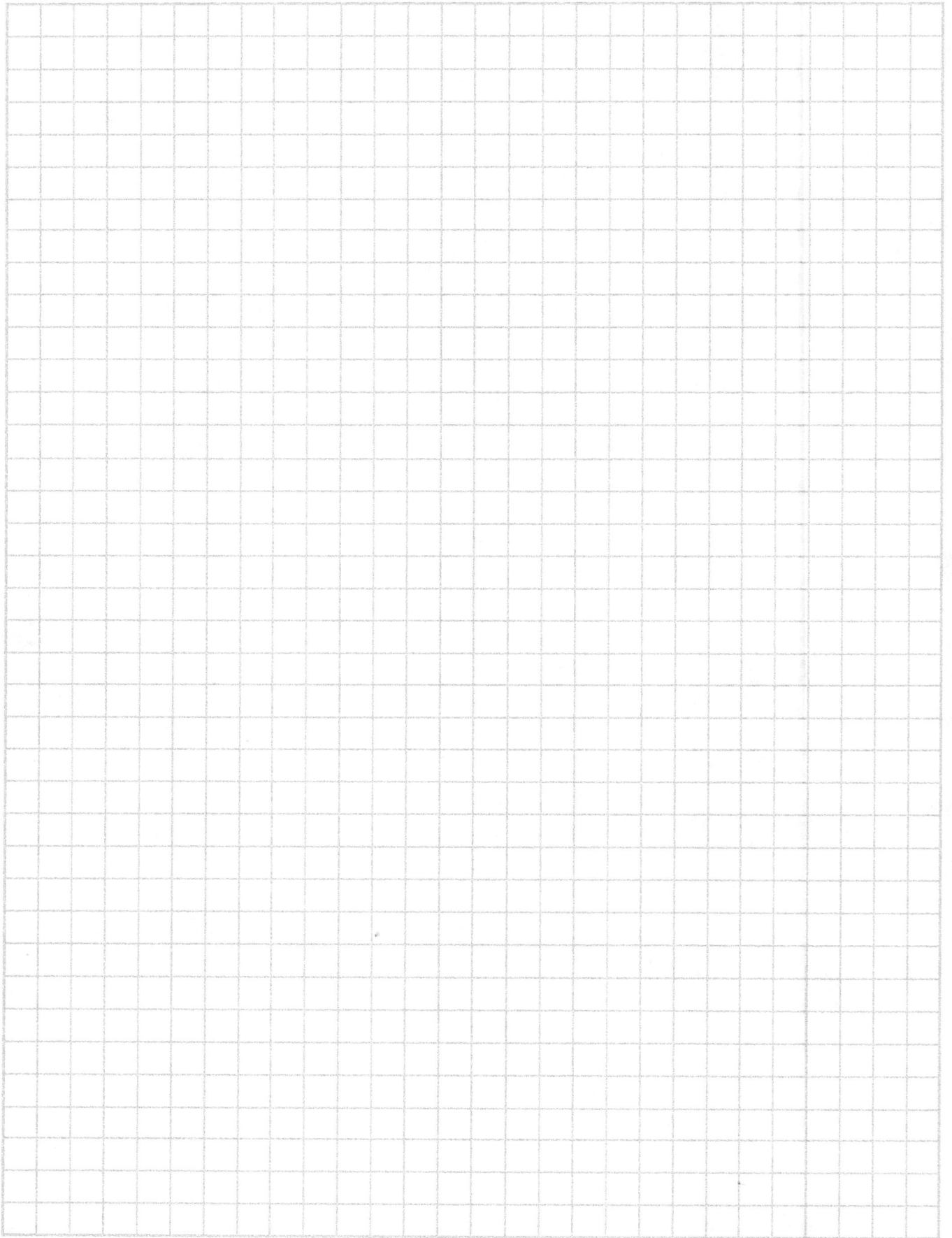

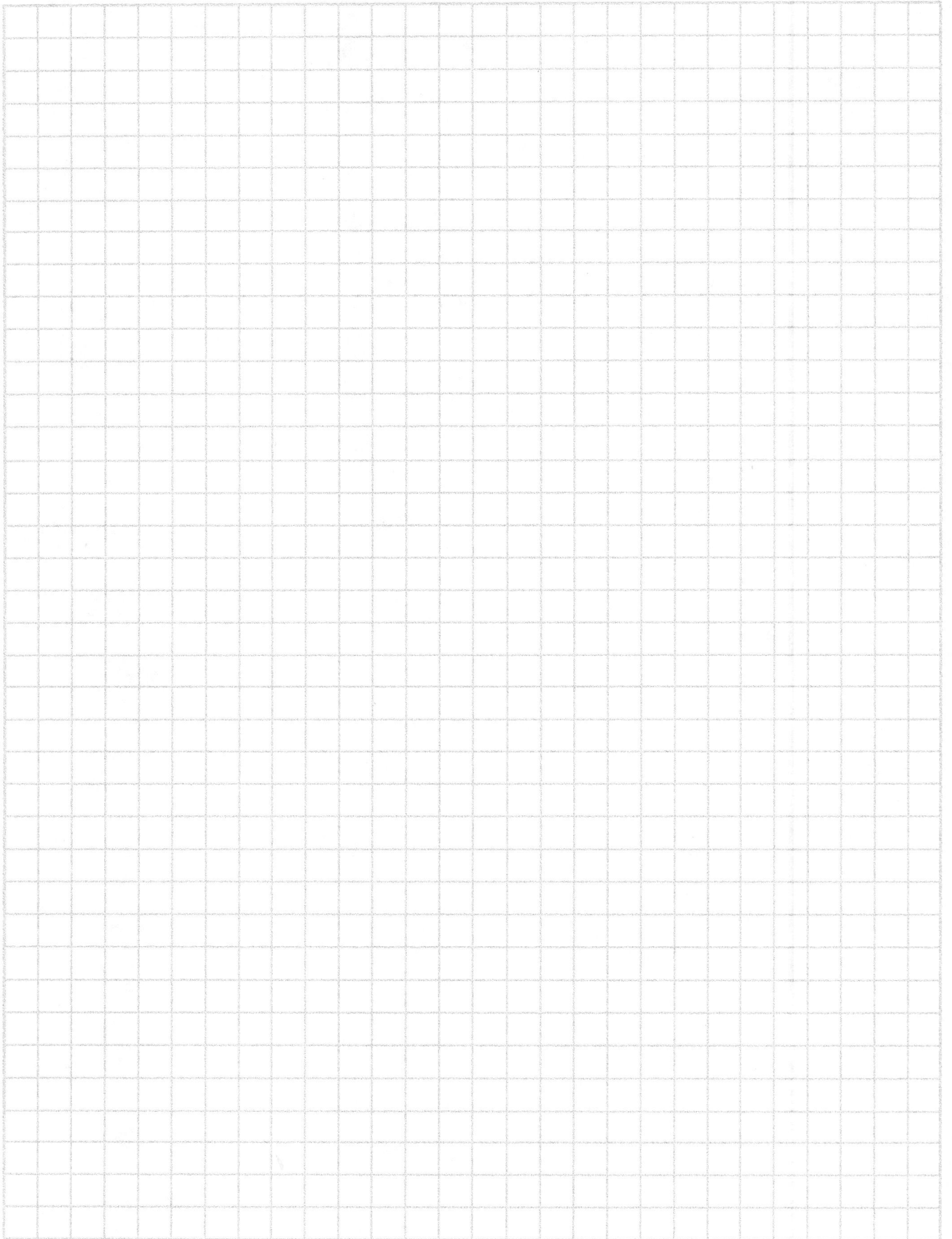

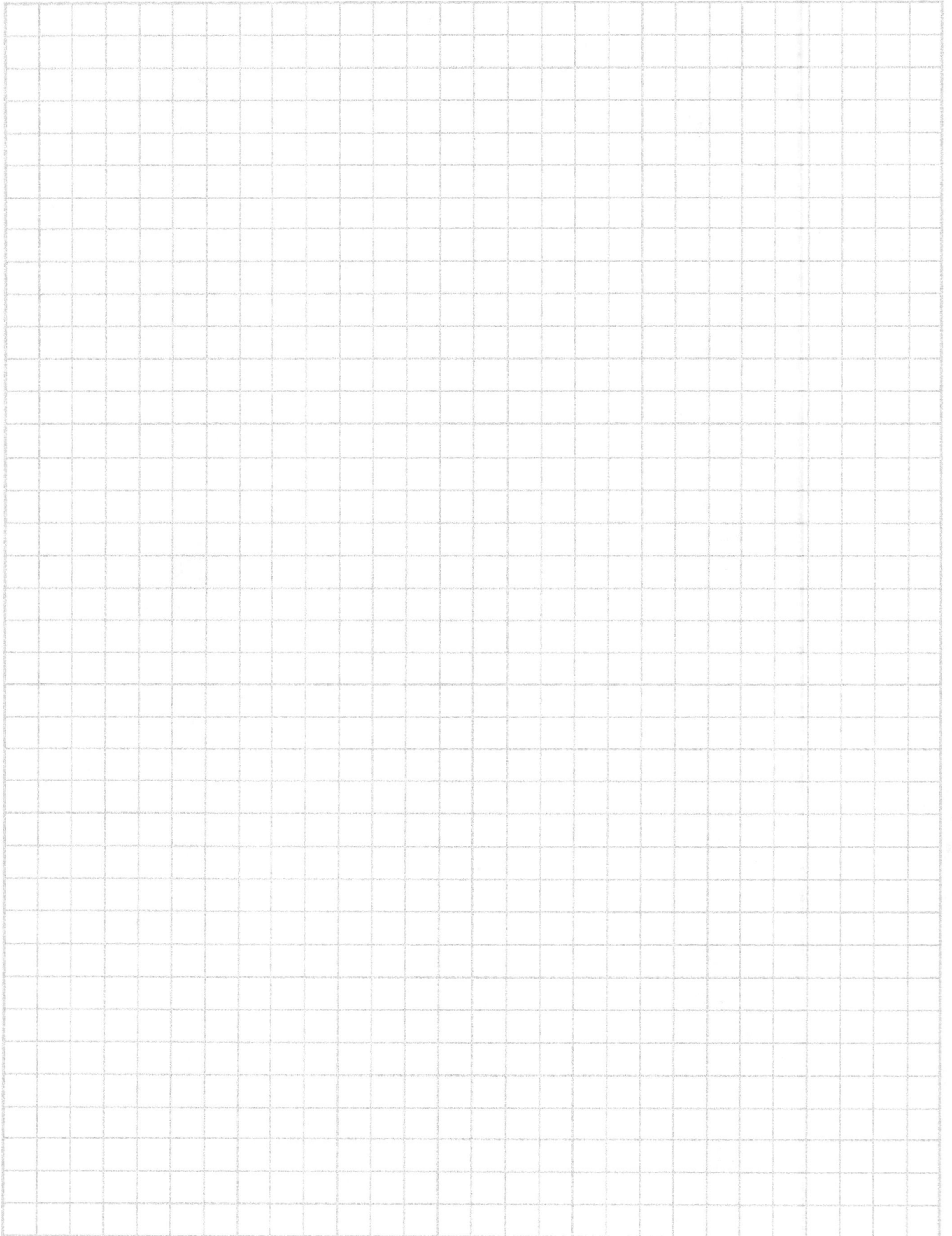

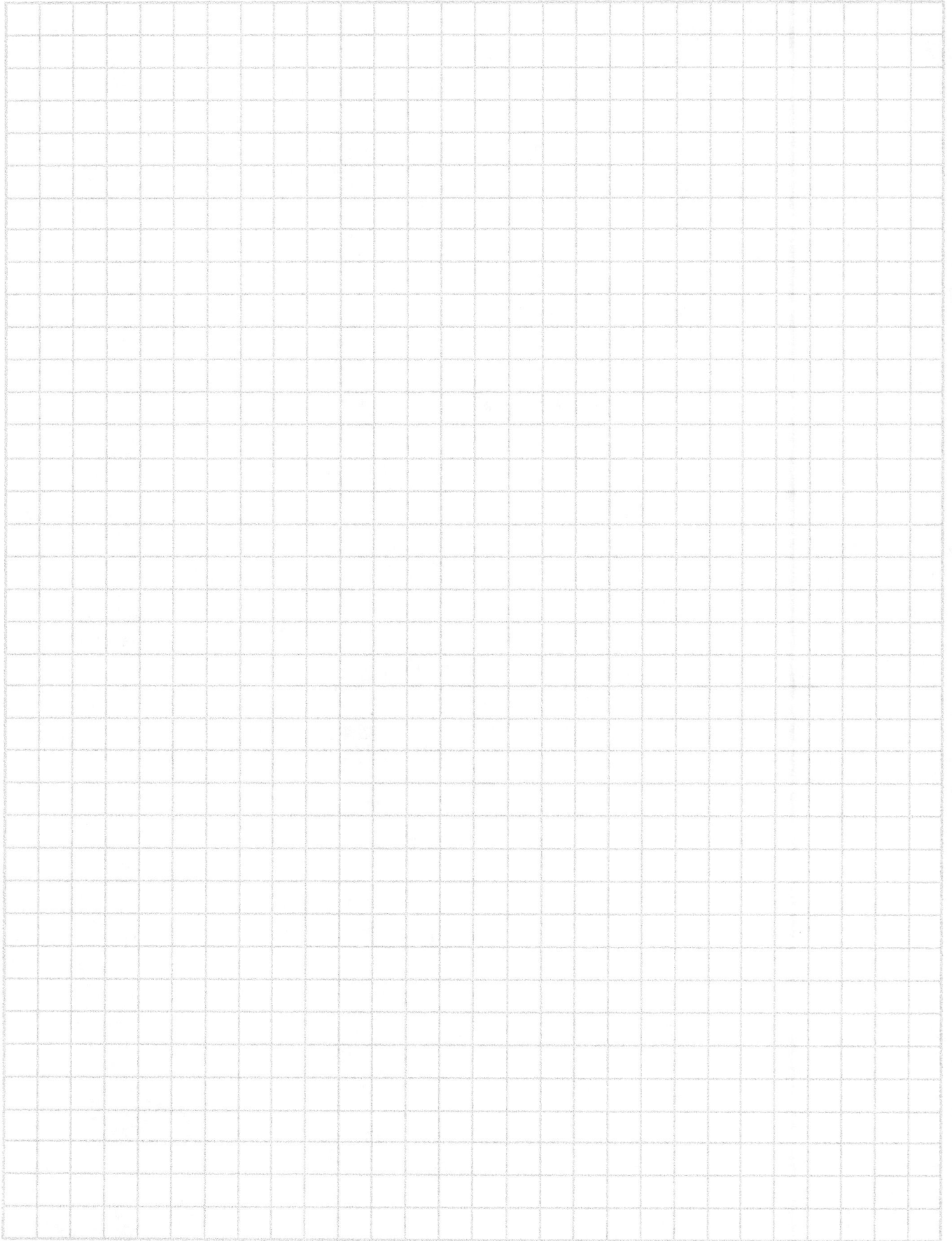

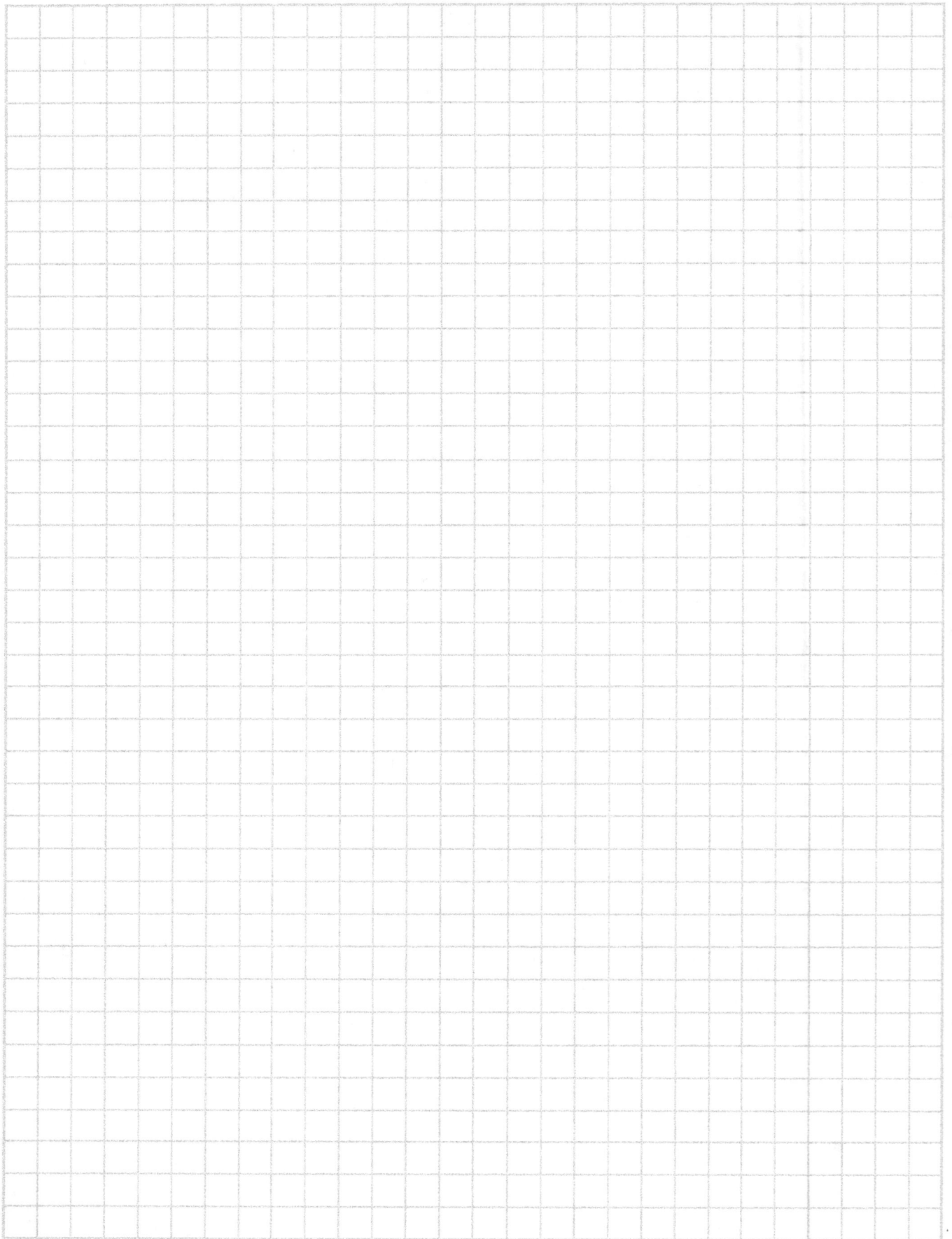

24

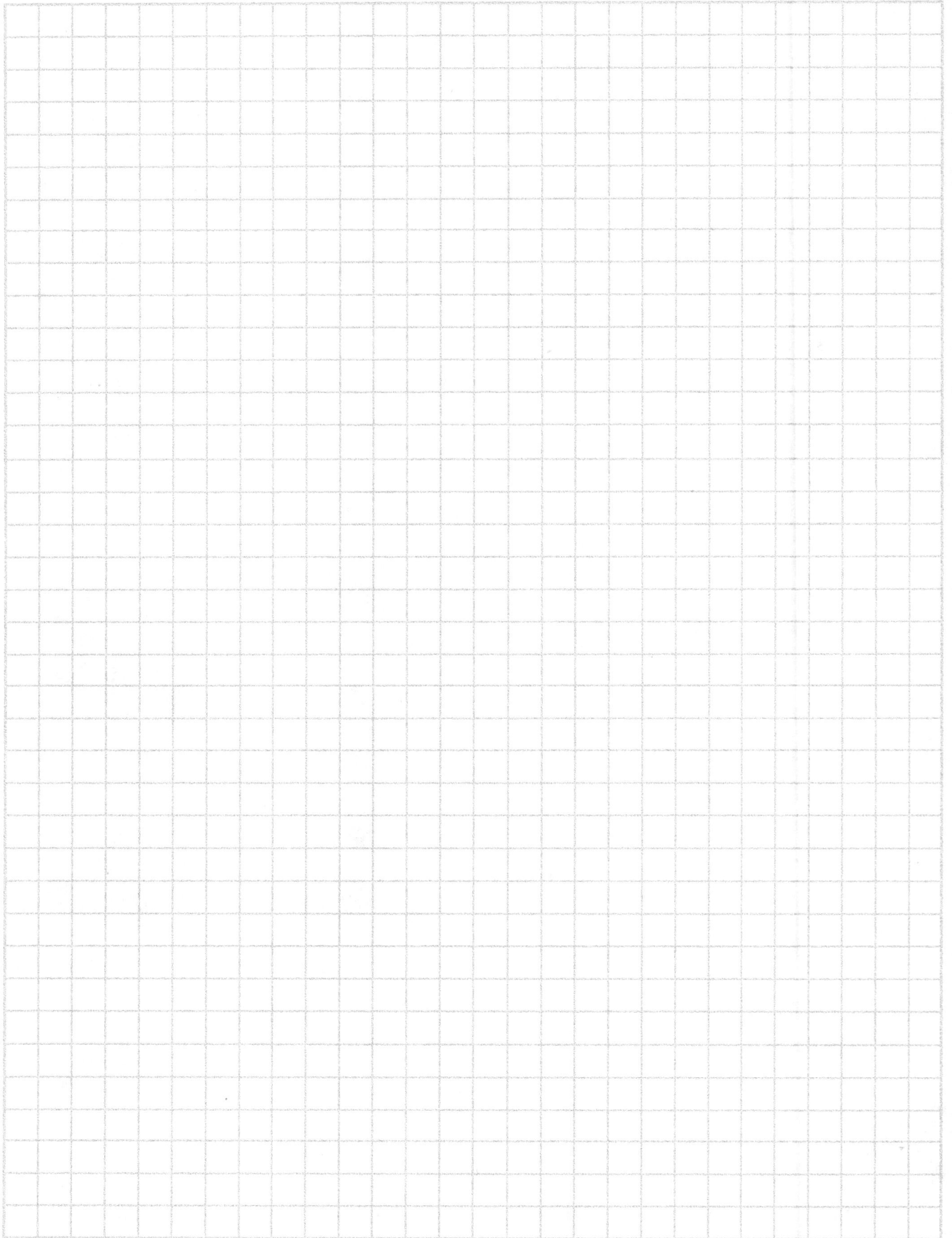

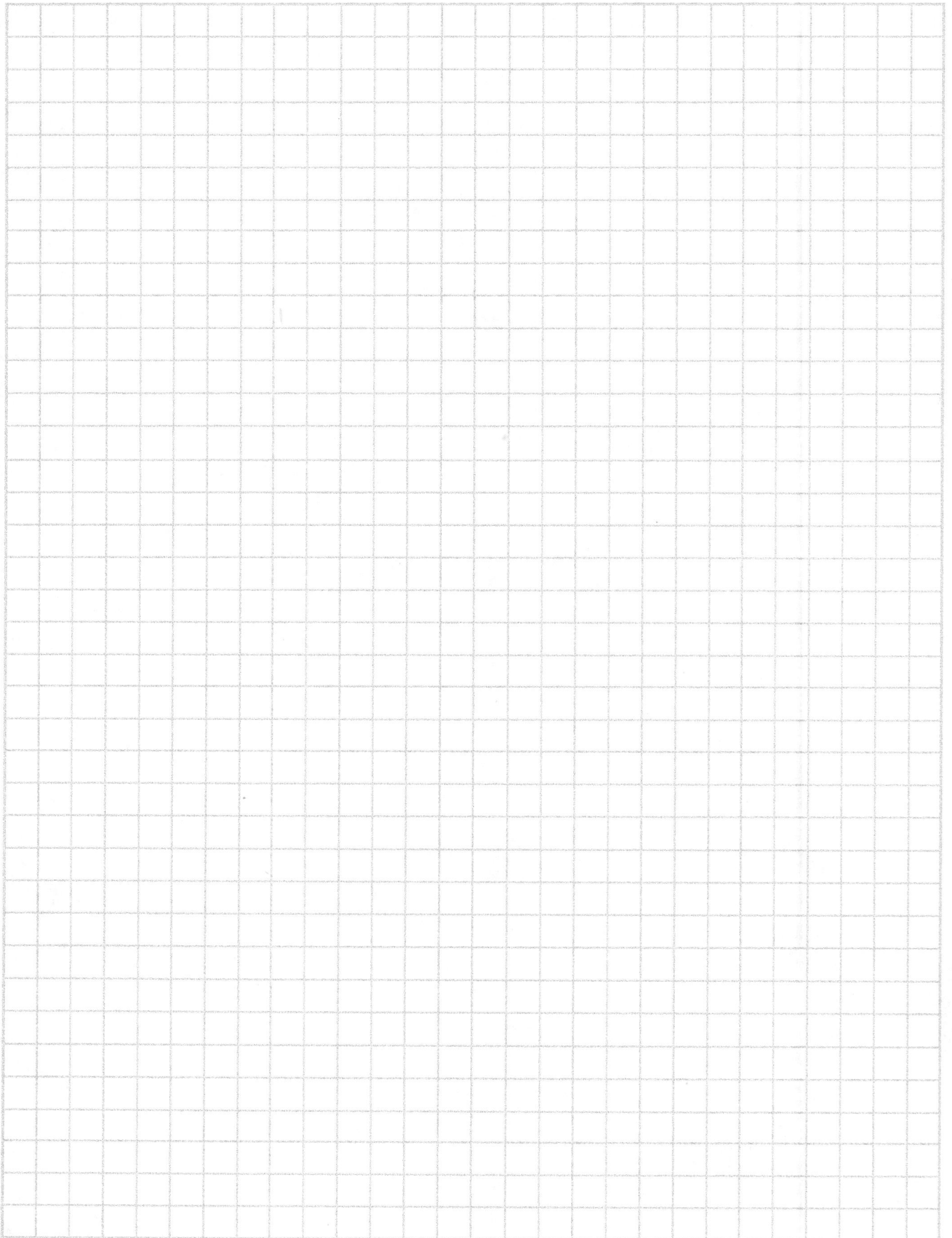

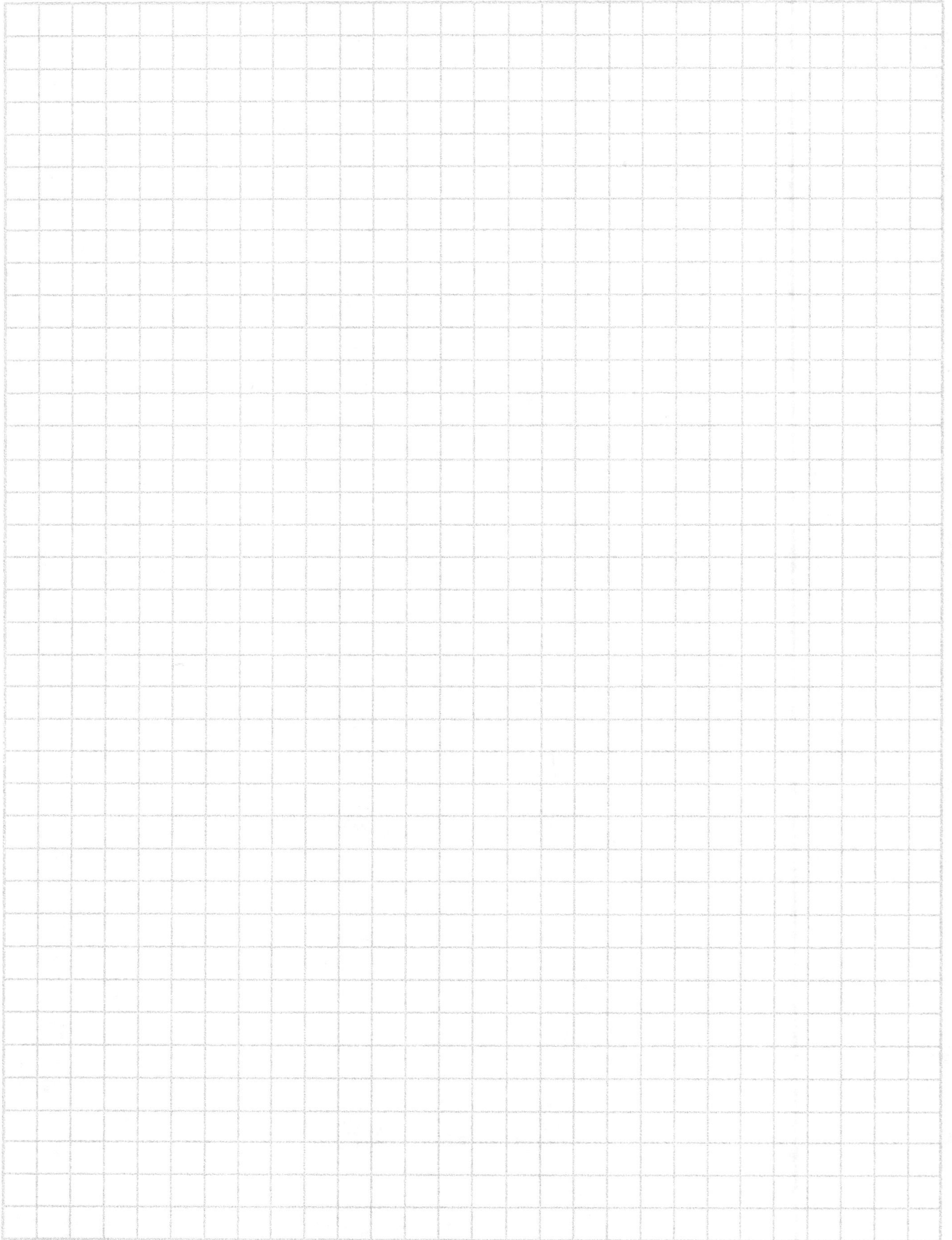

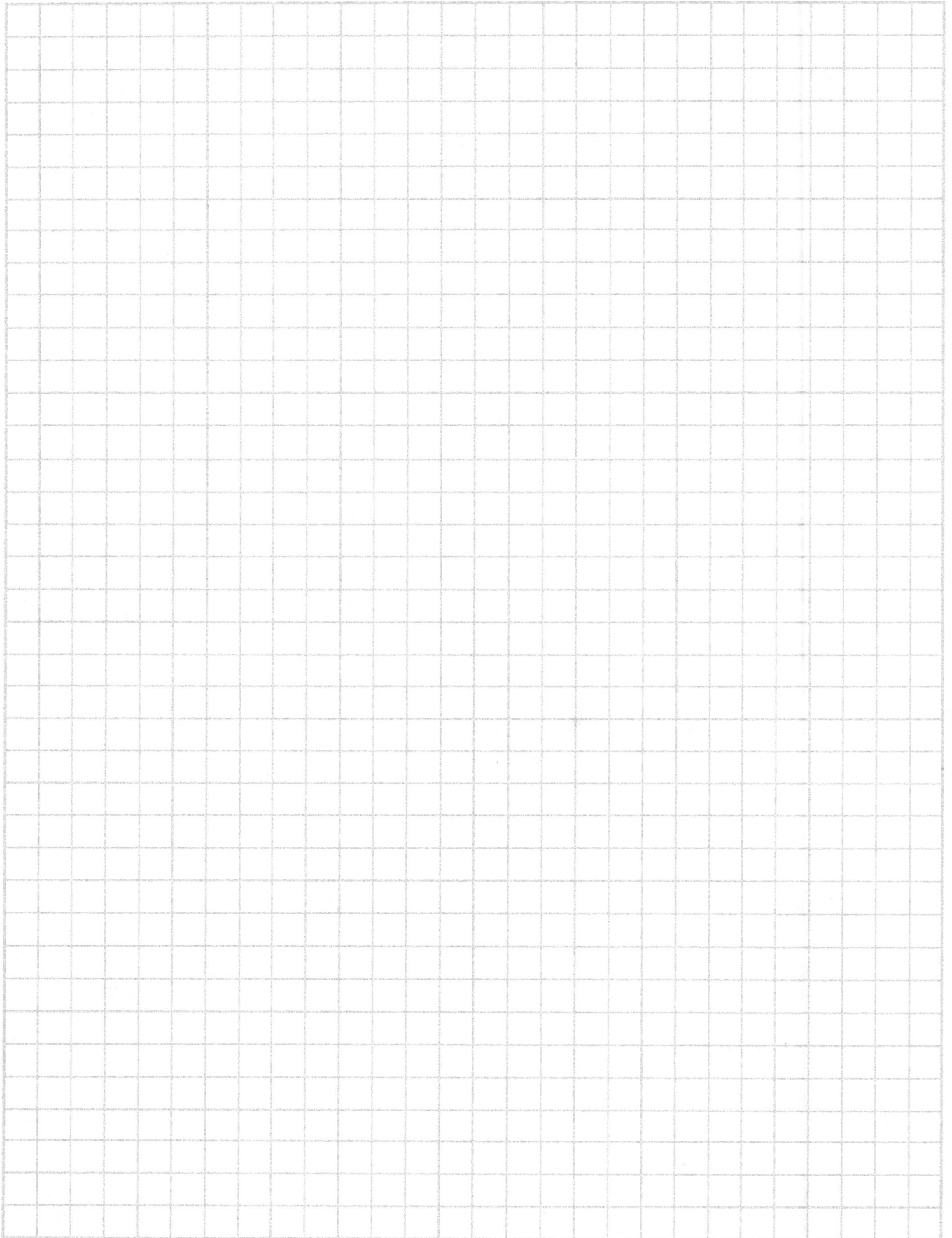

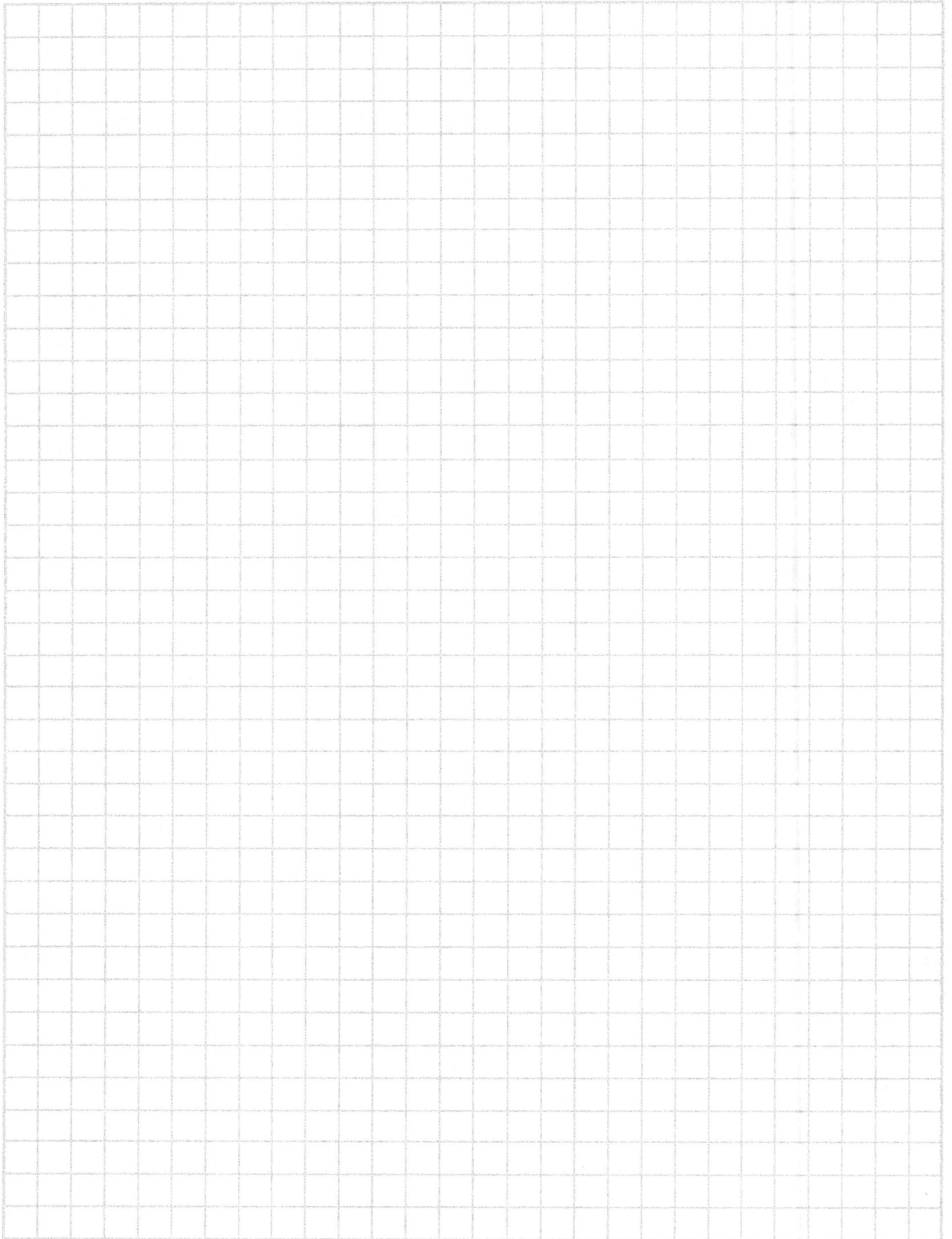

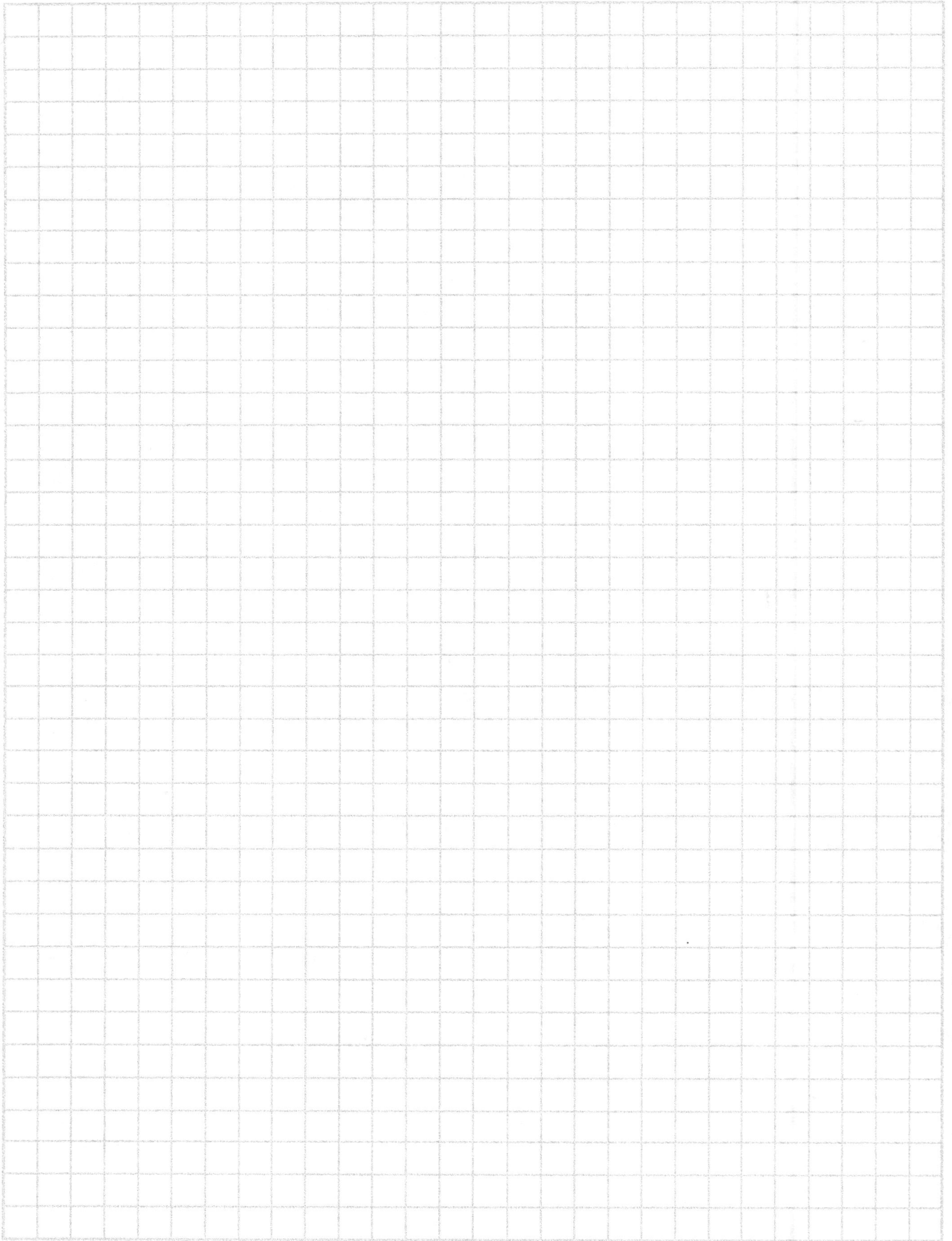

38

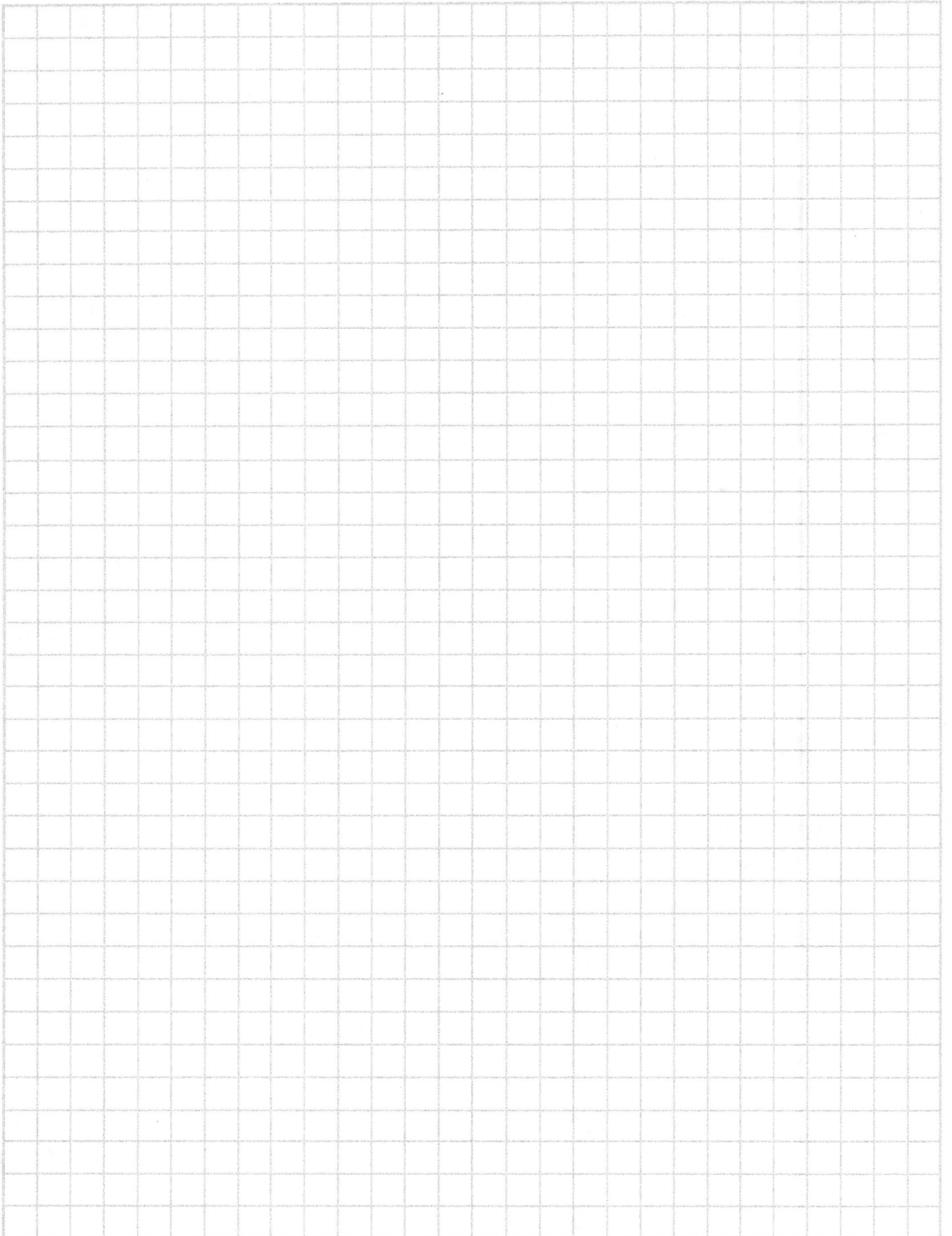

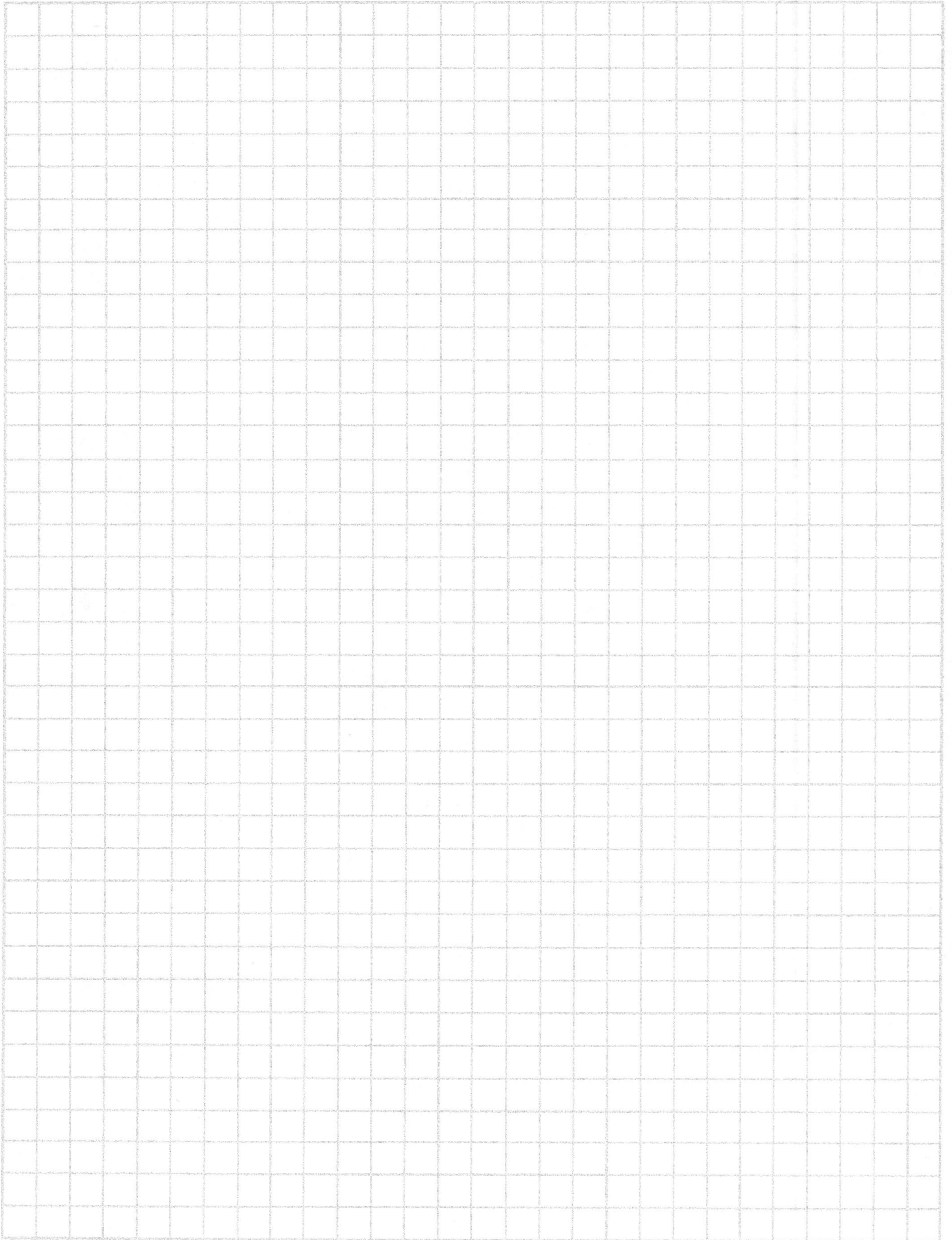

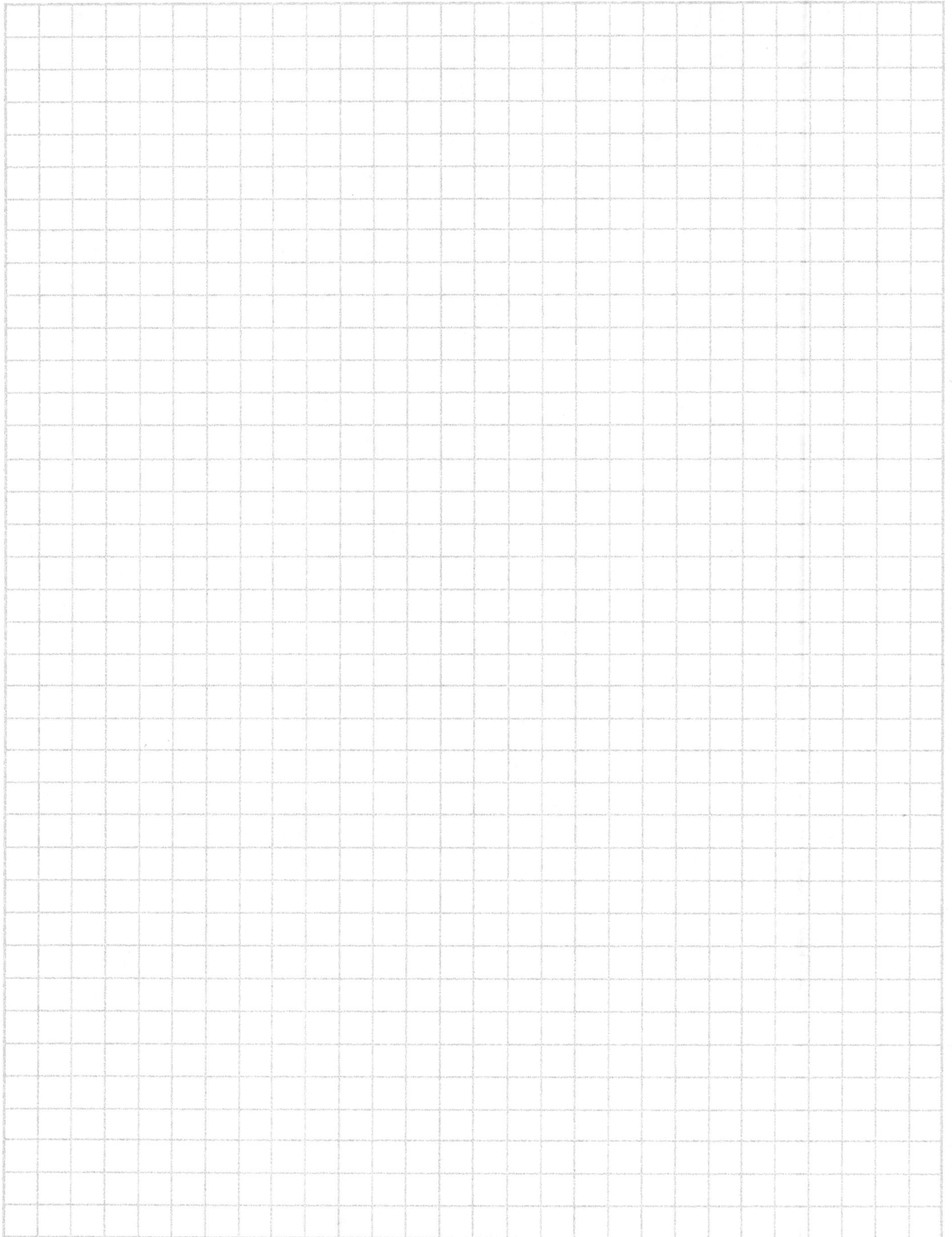

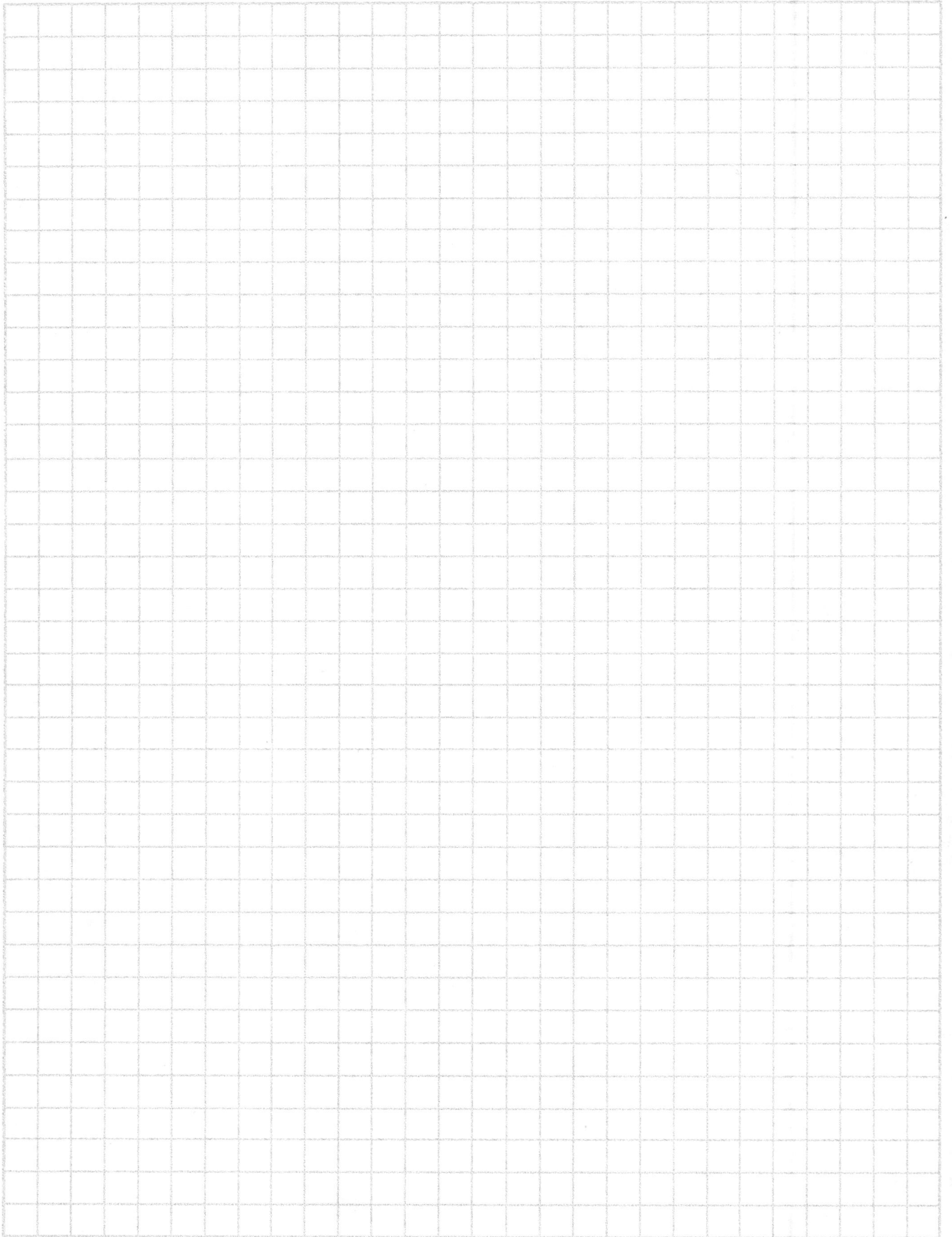

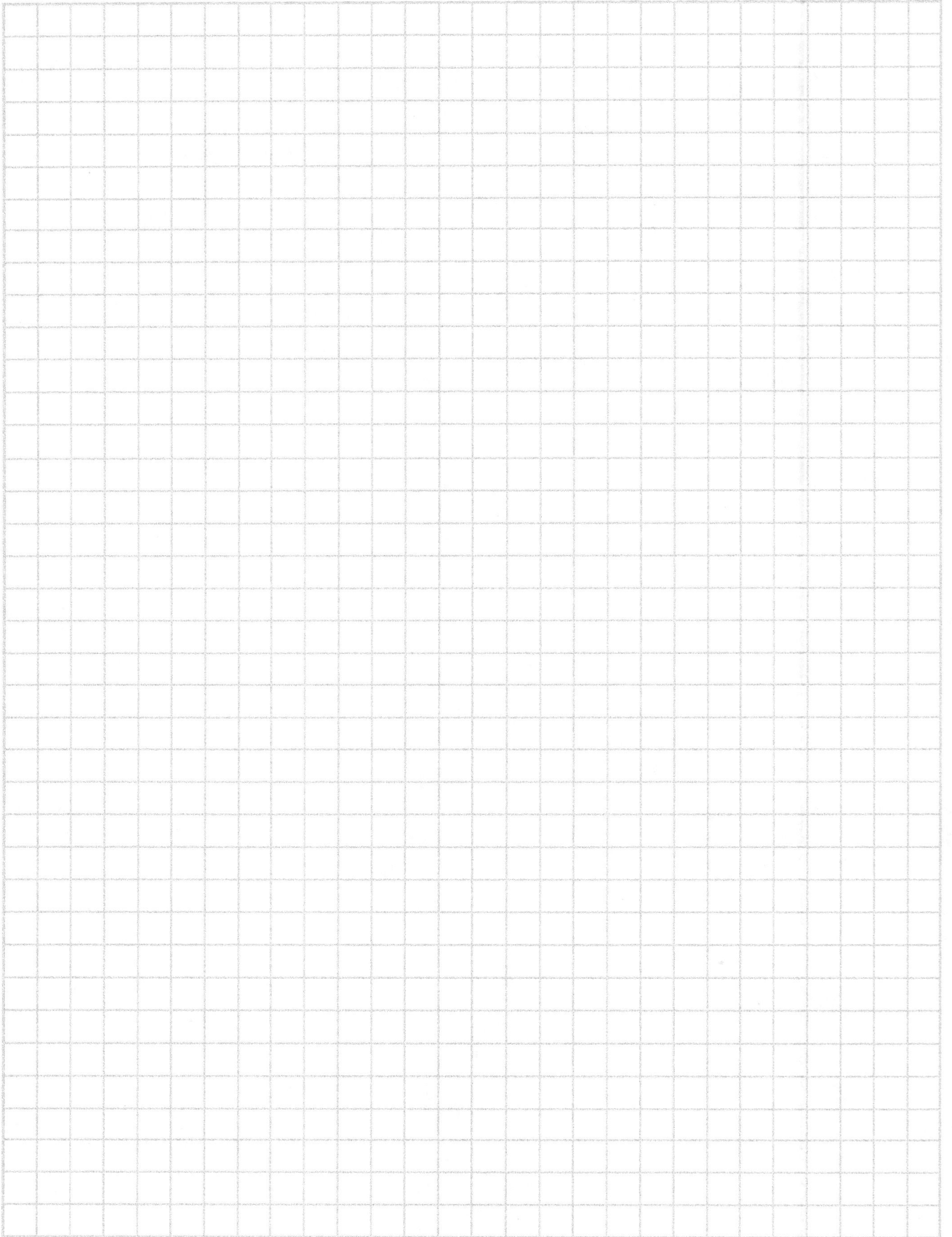

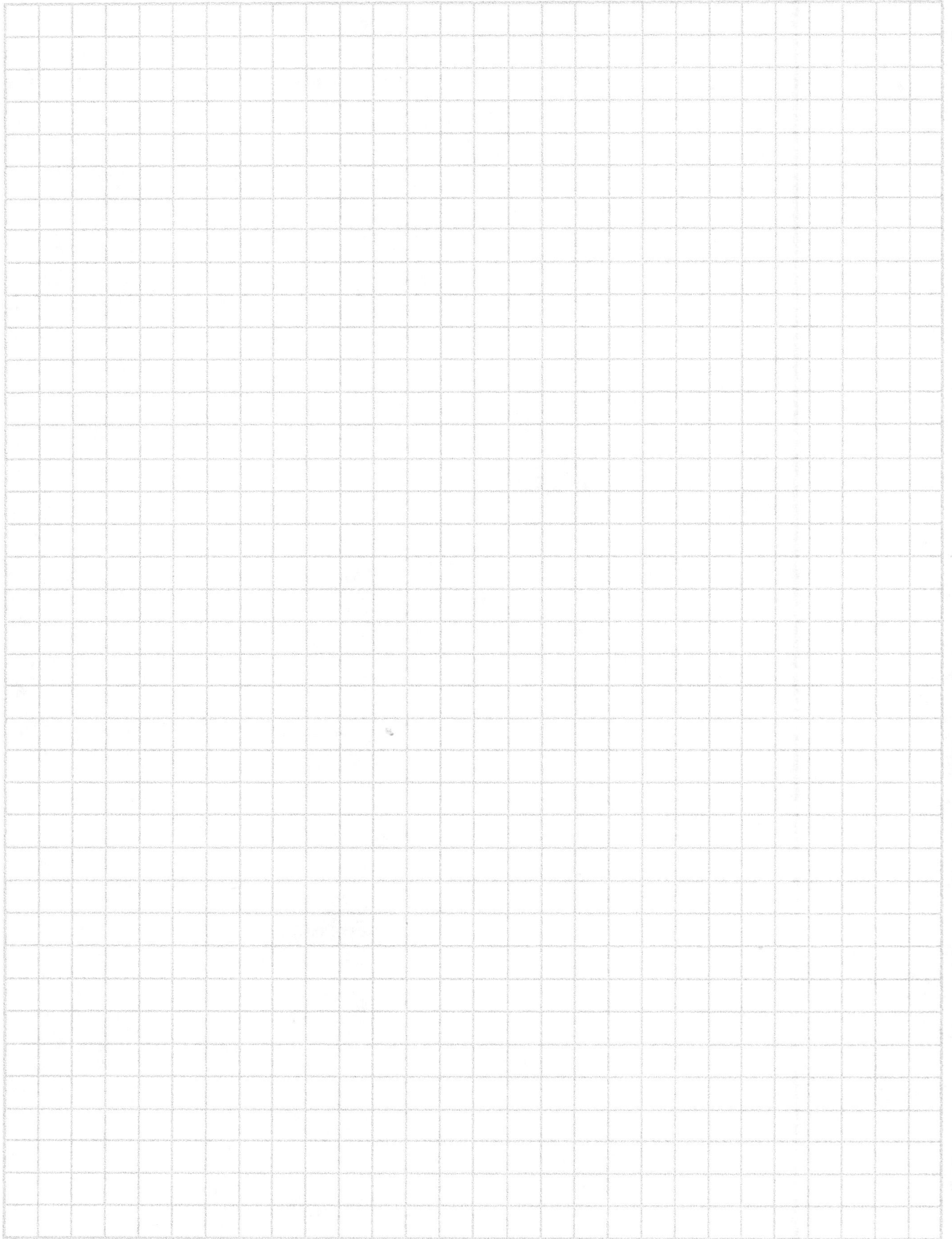

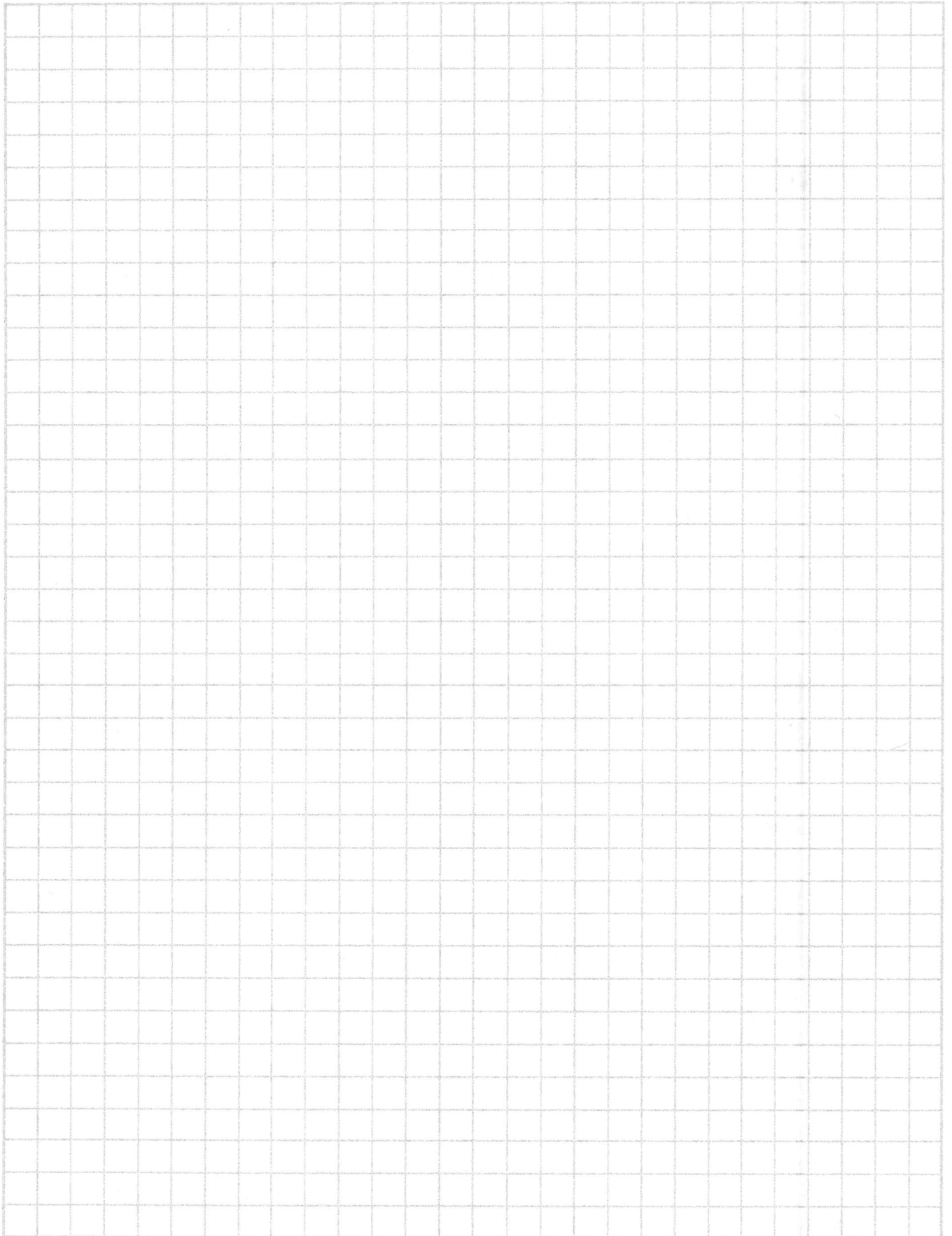

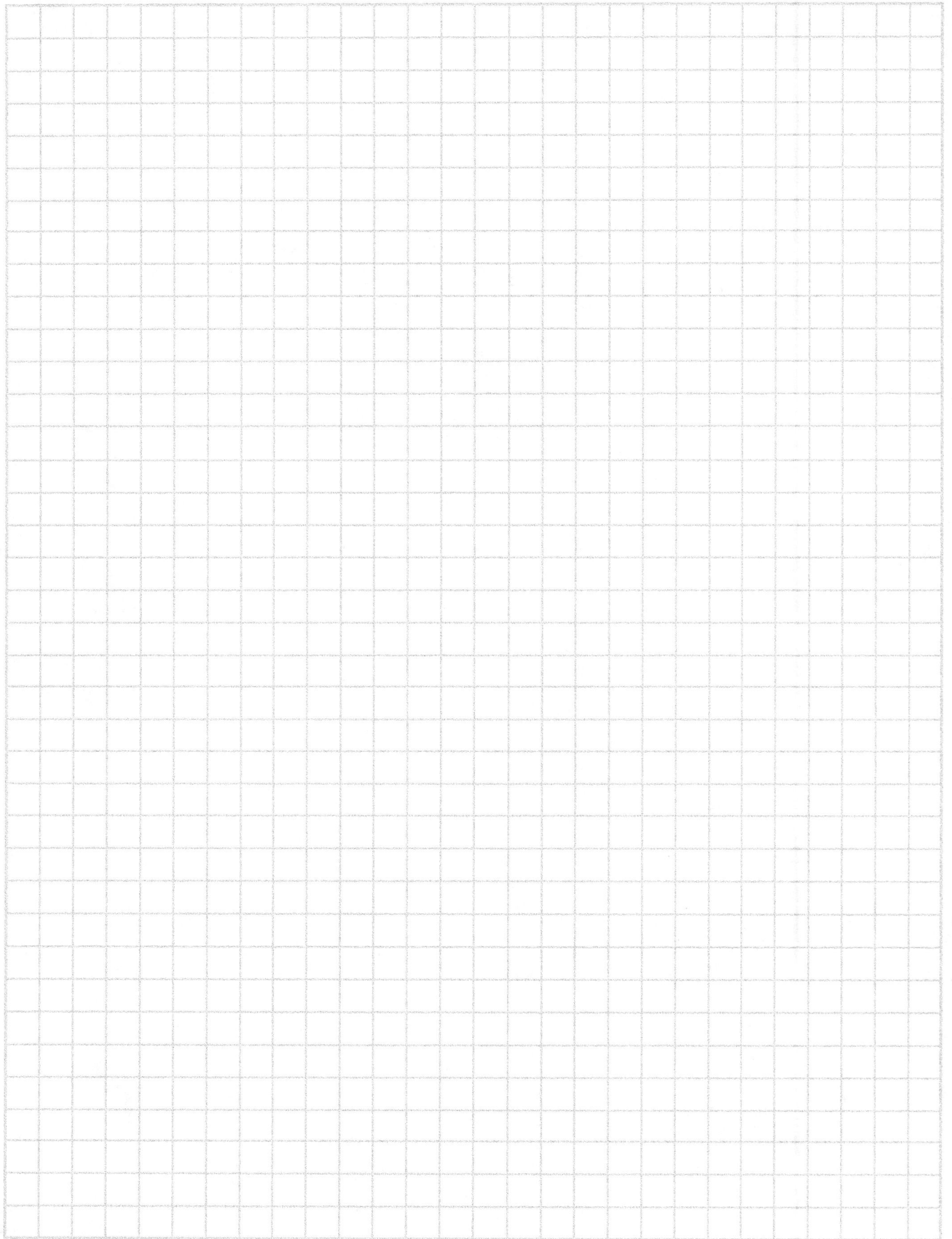

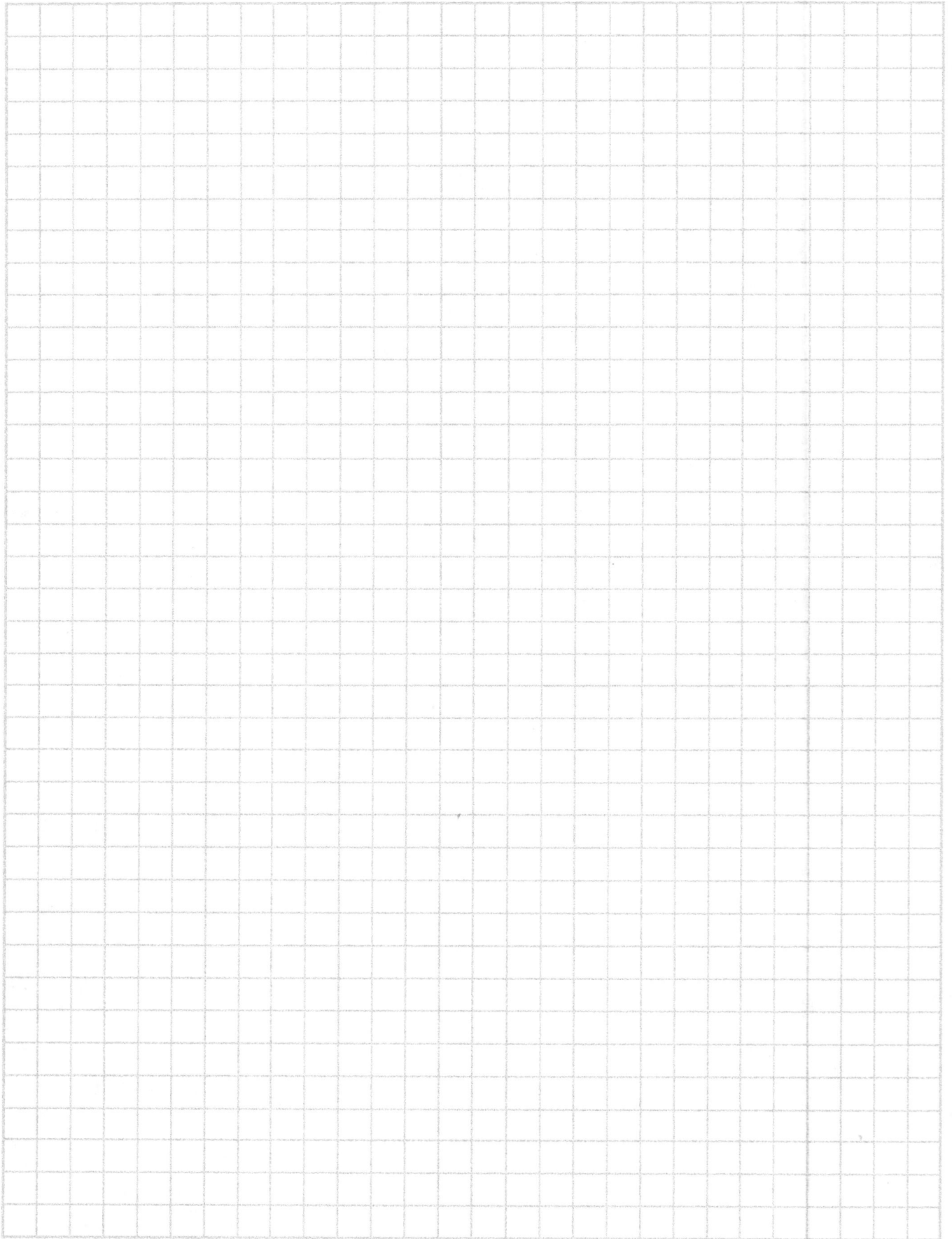

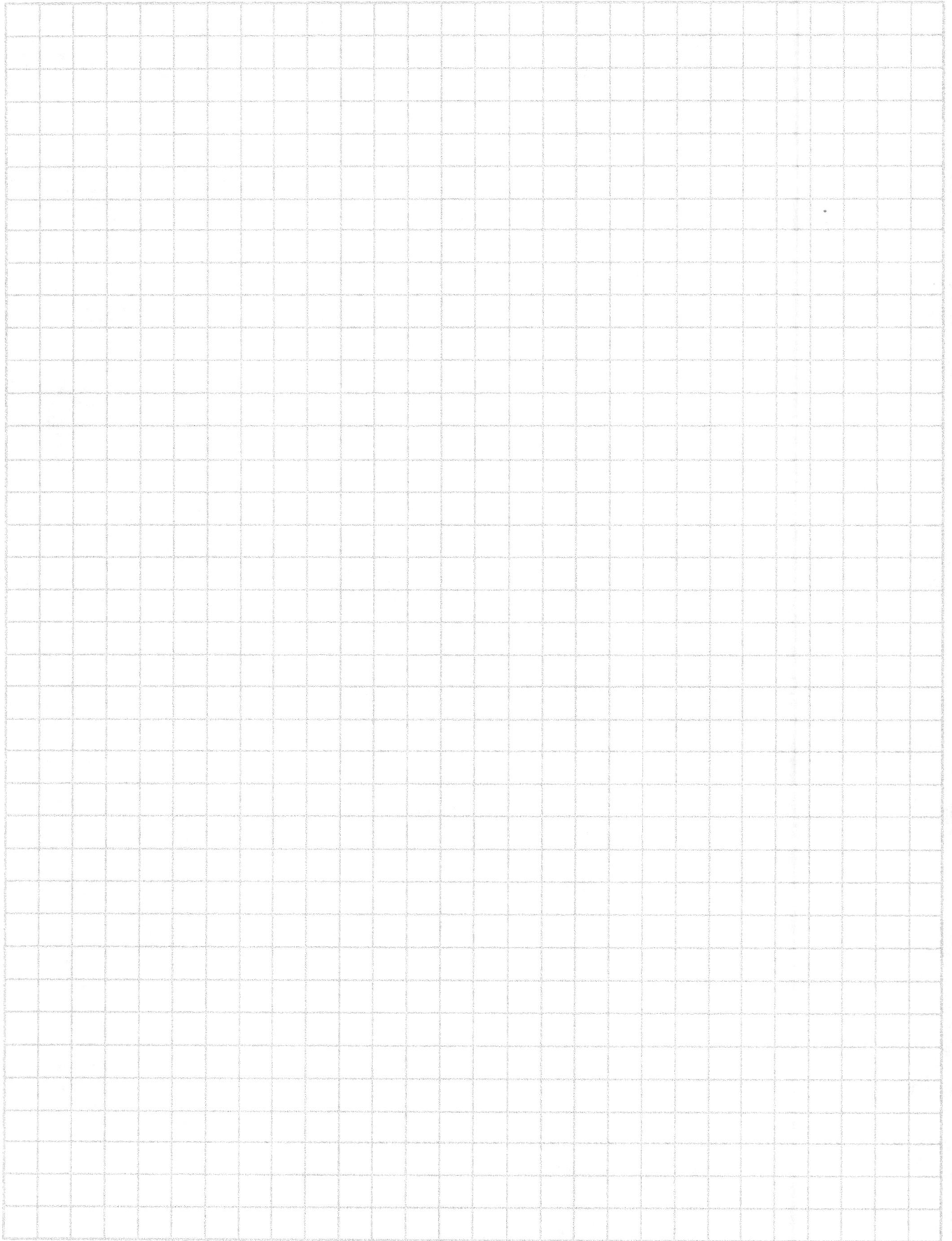

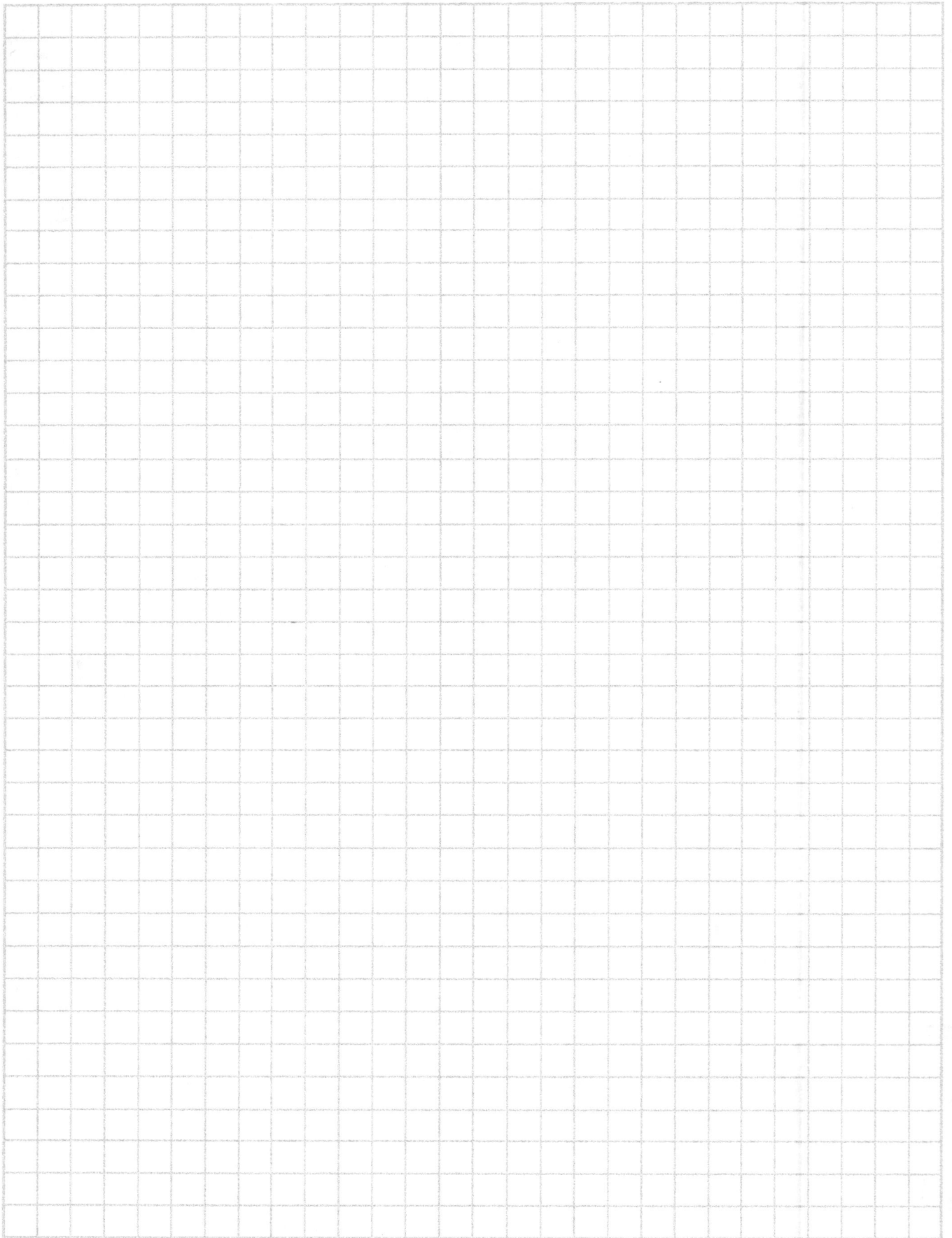

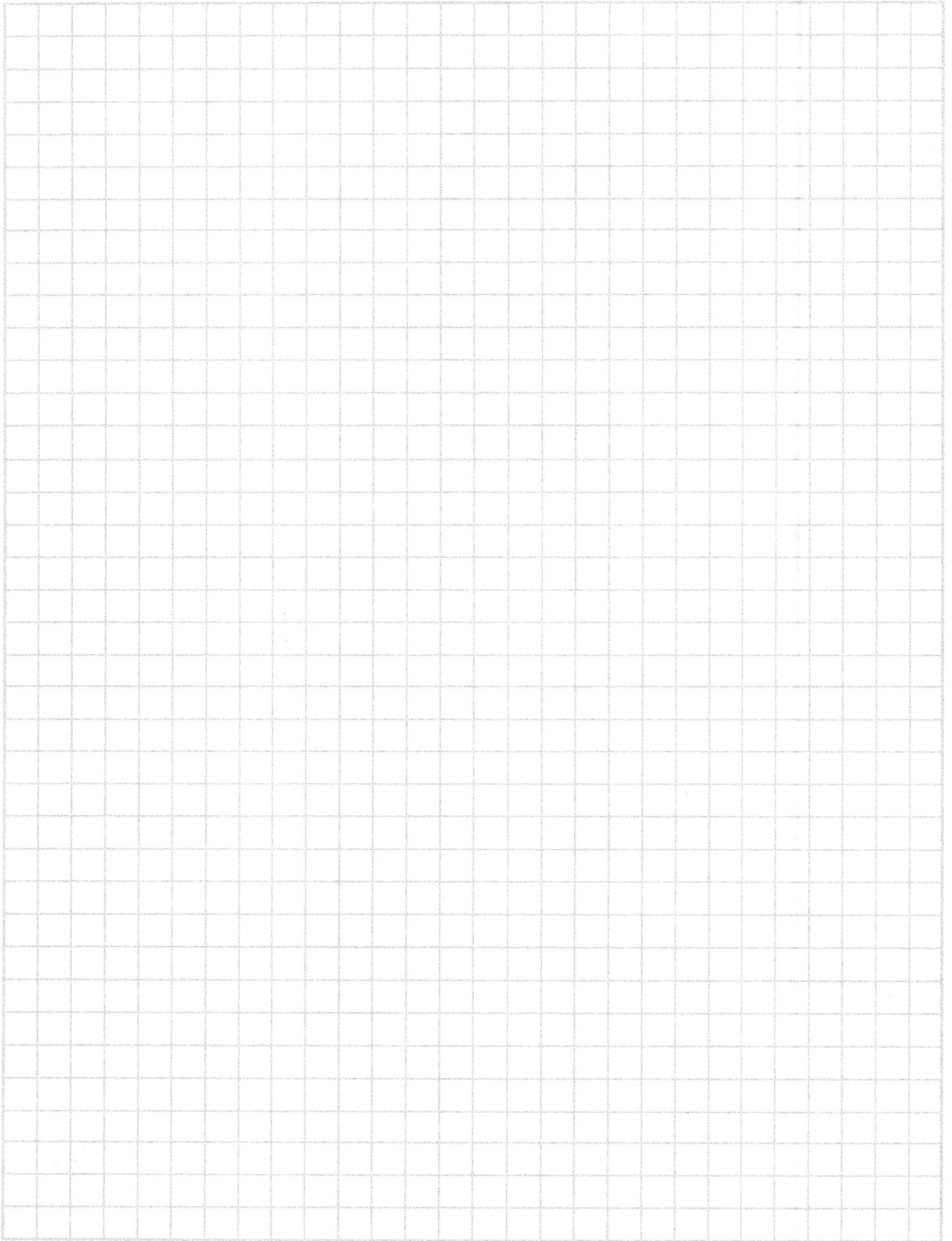

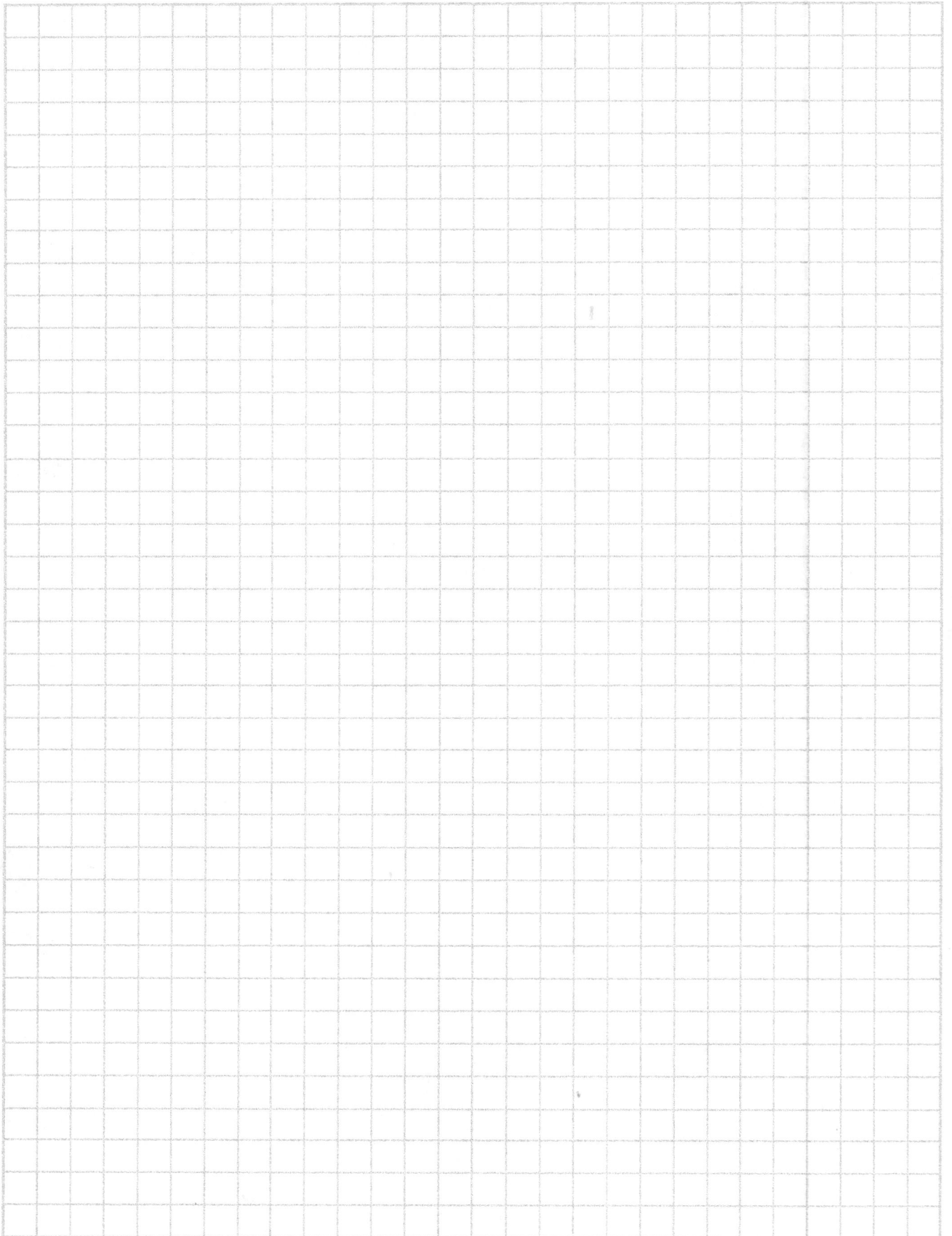

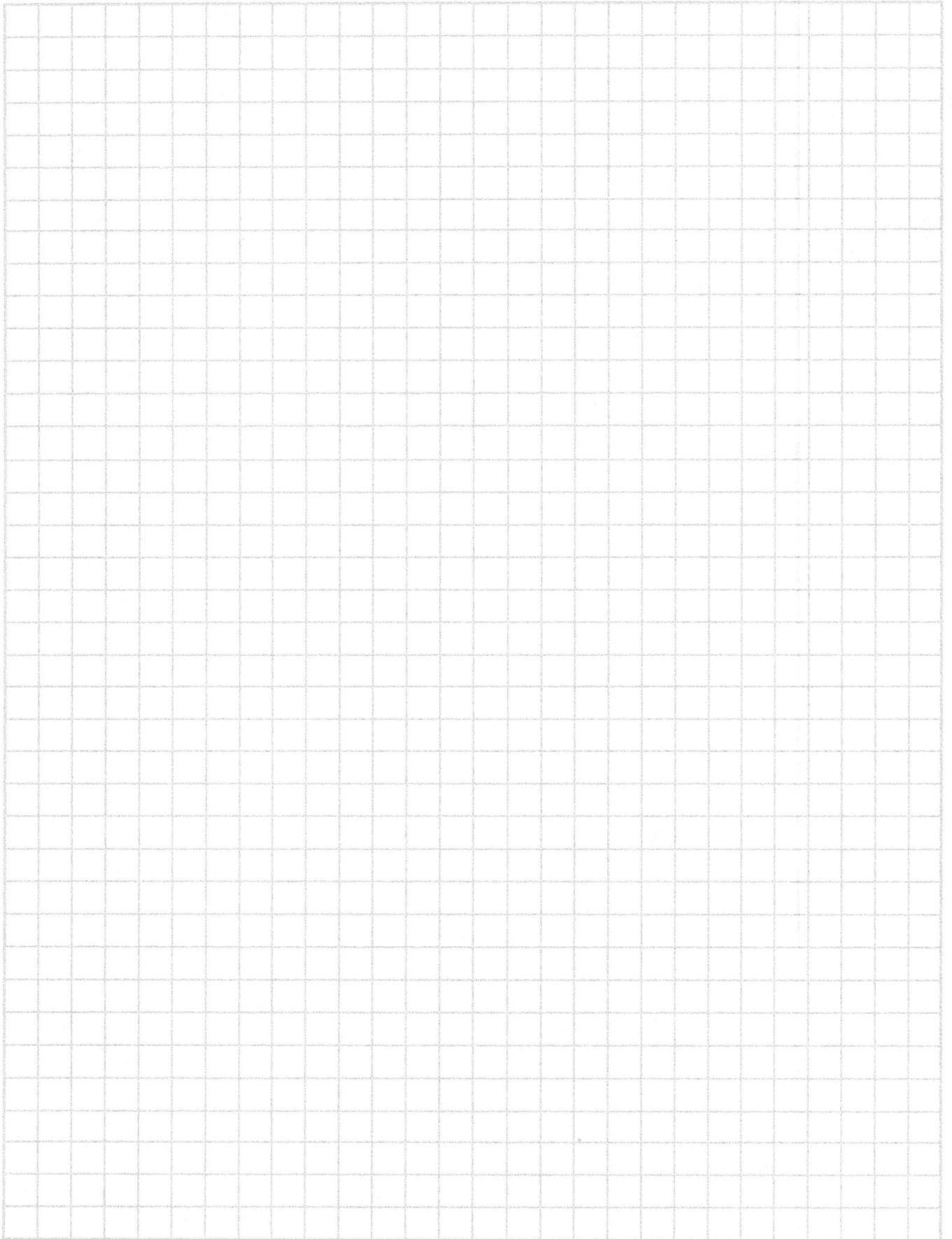

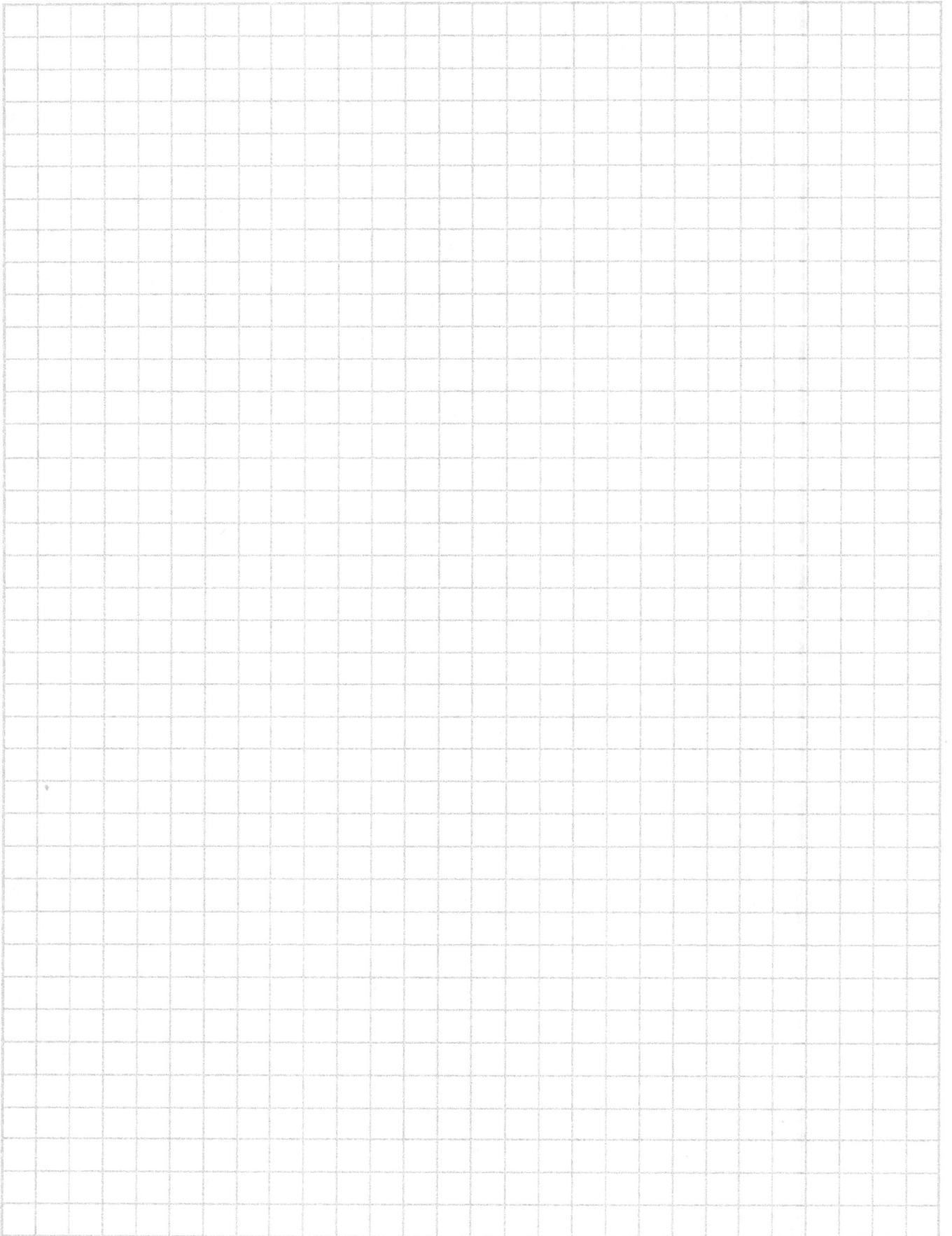

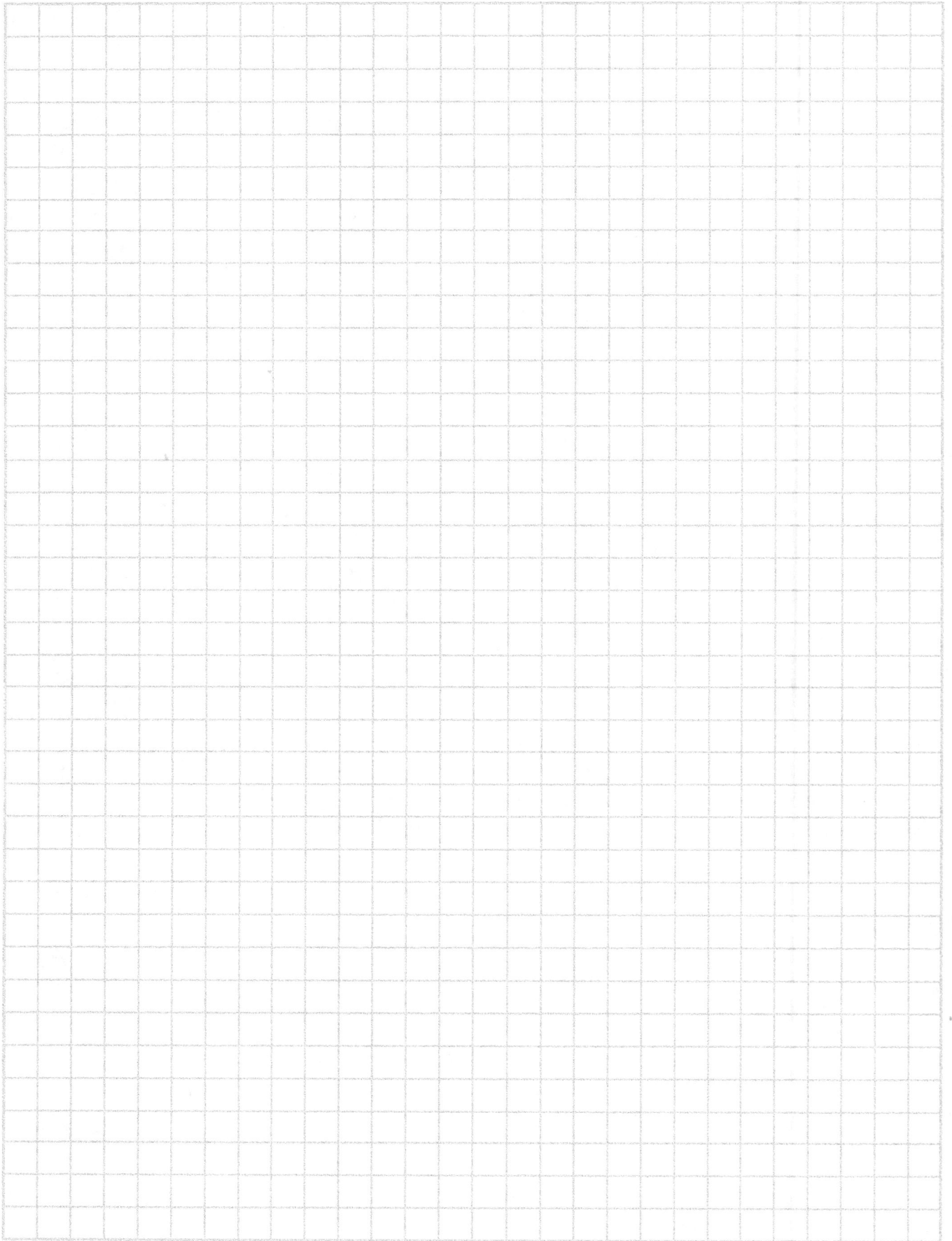

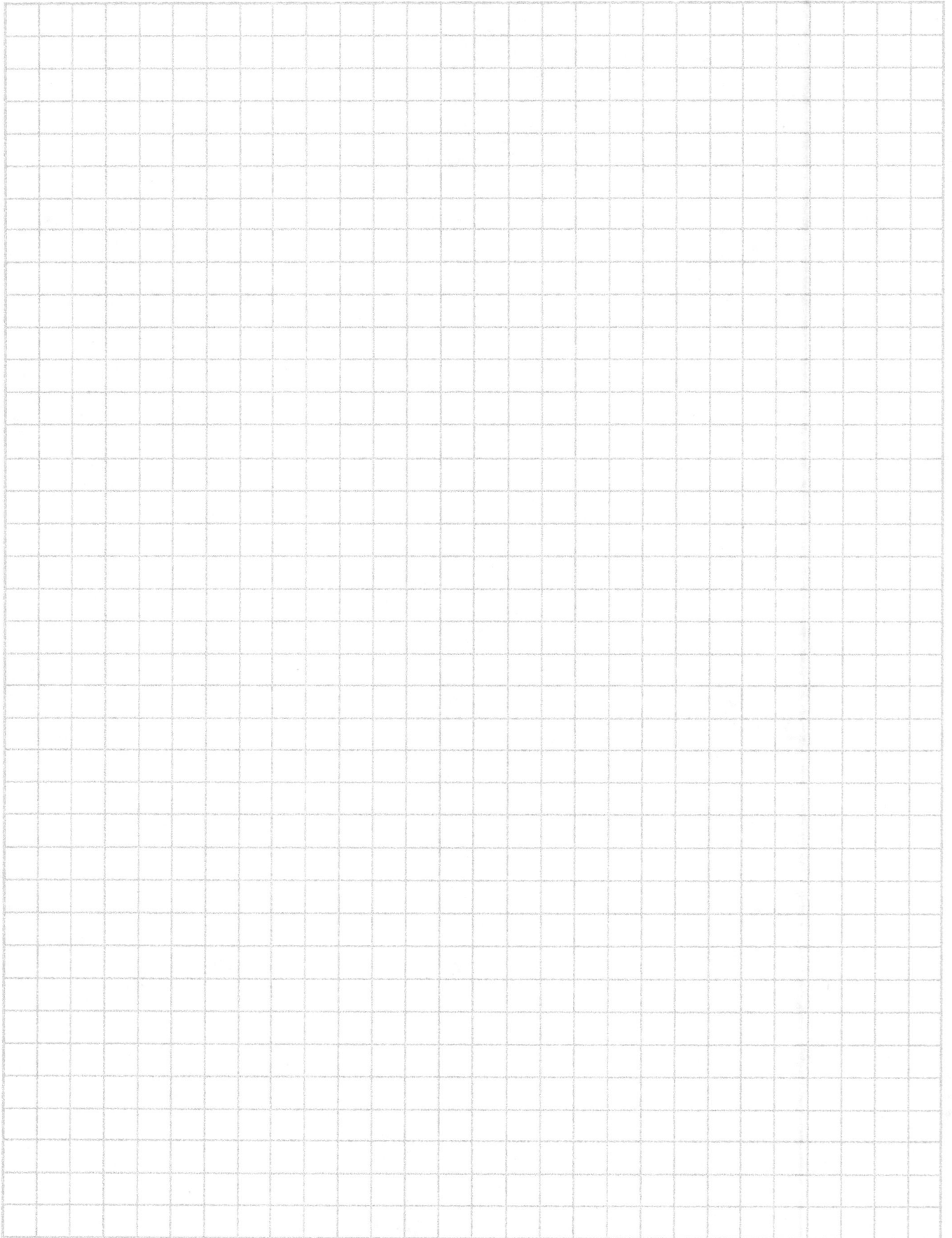

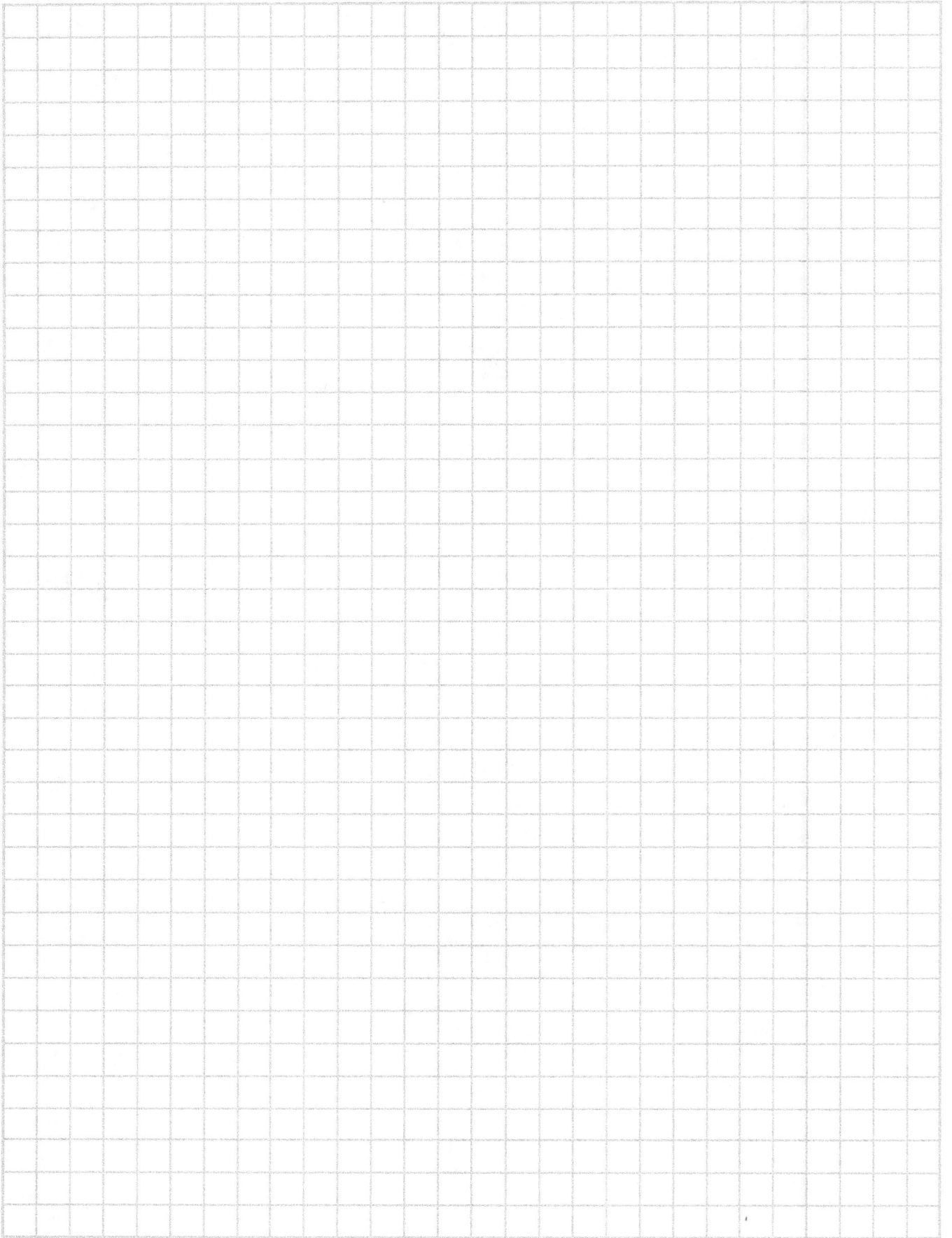

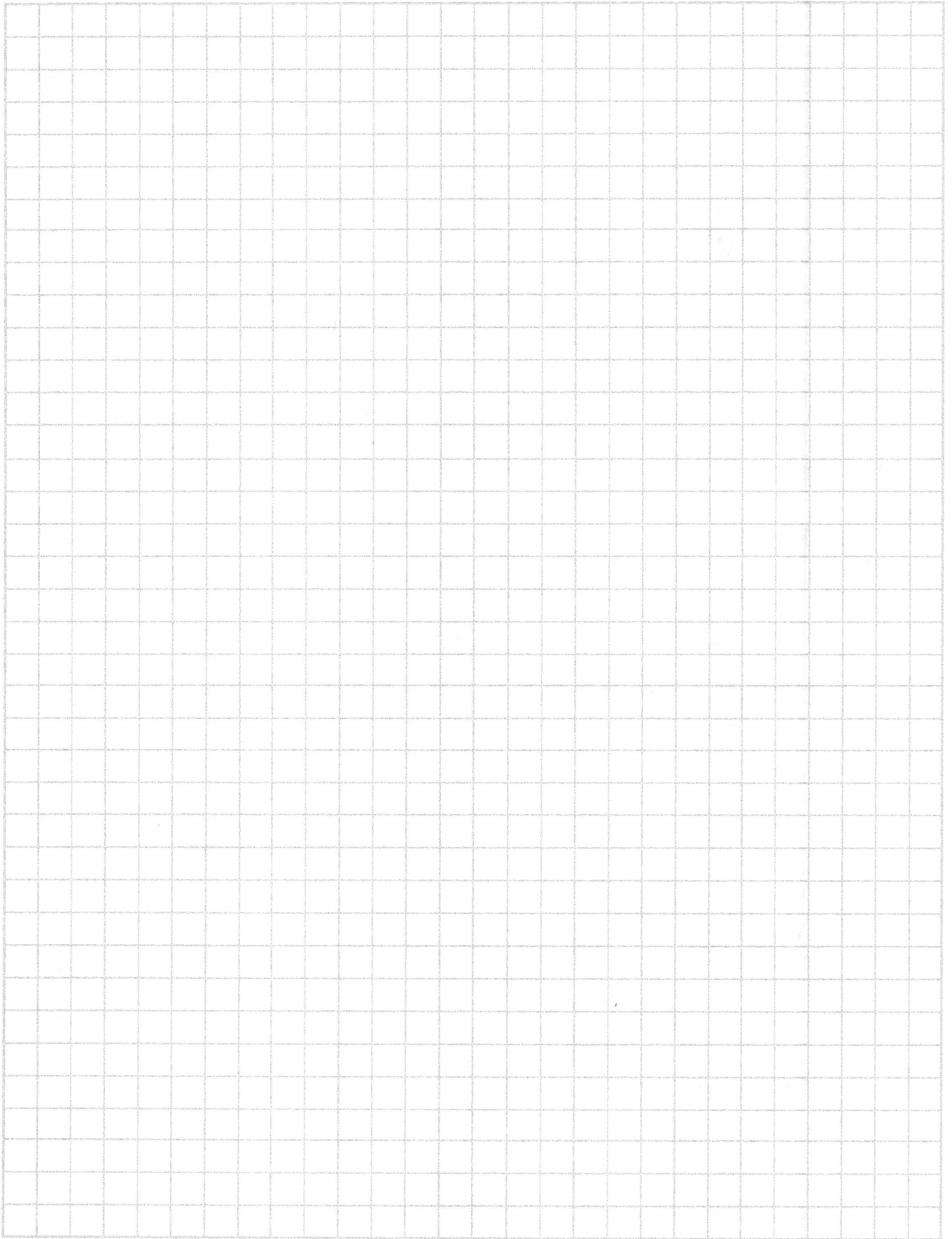

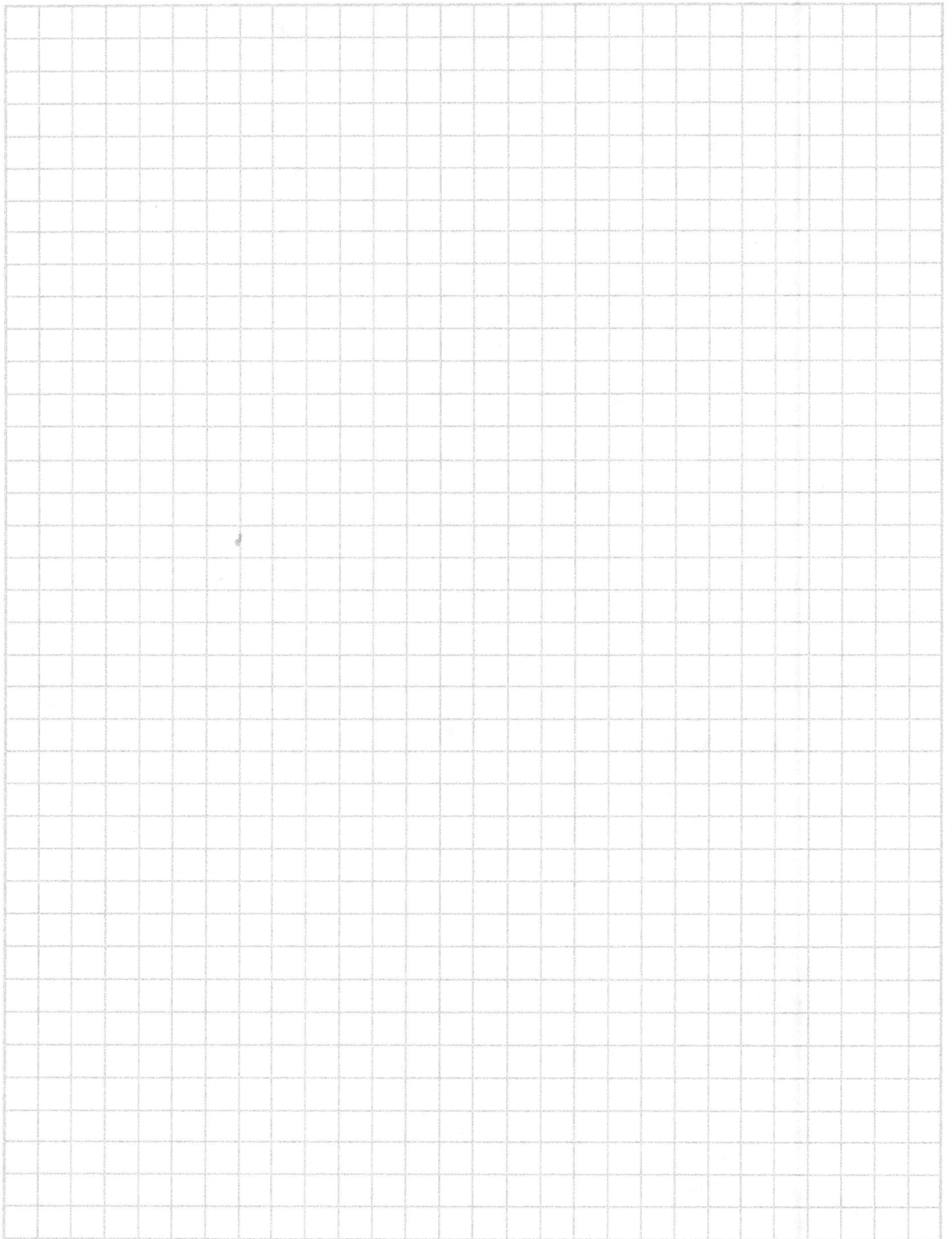

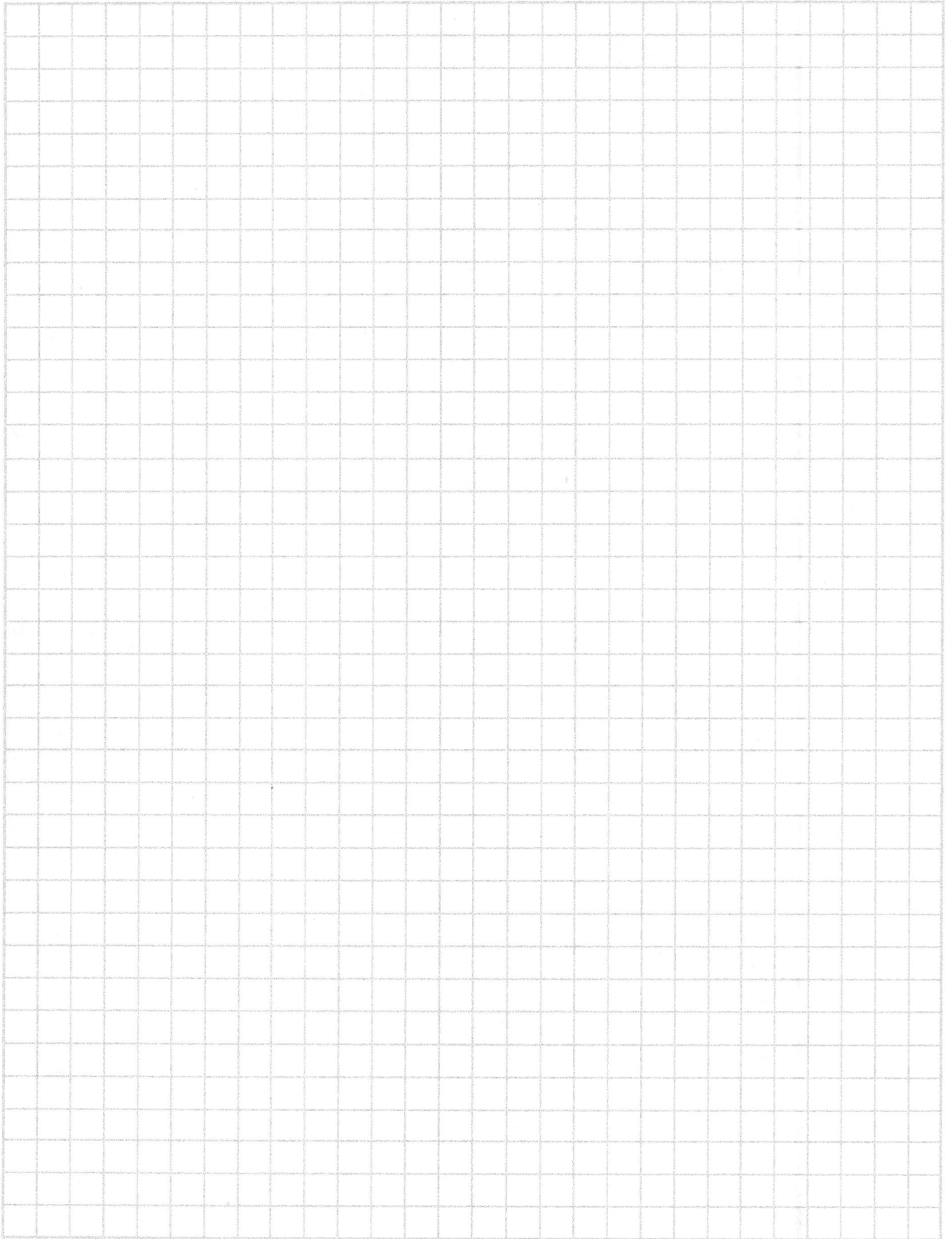

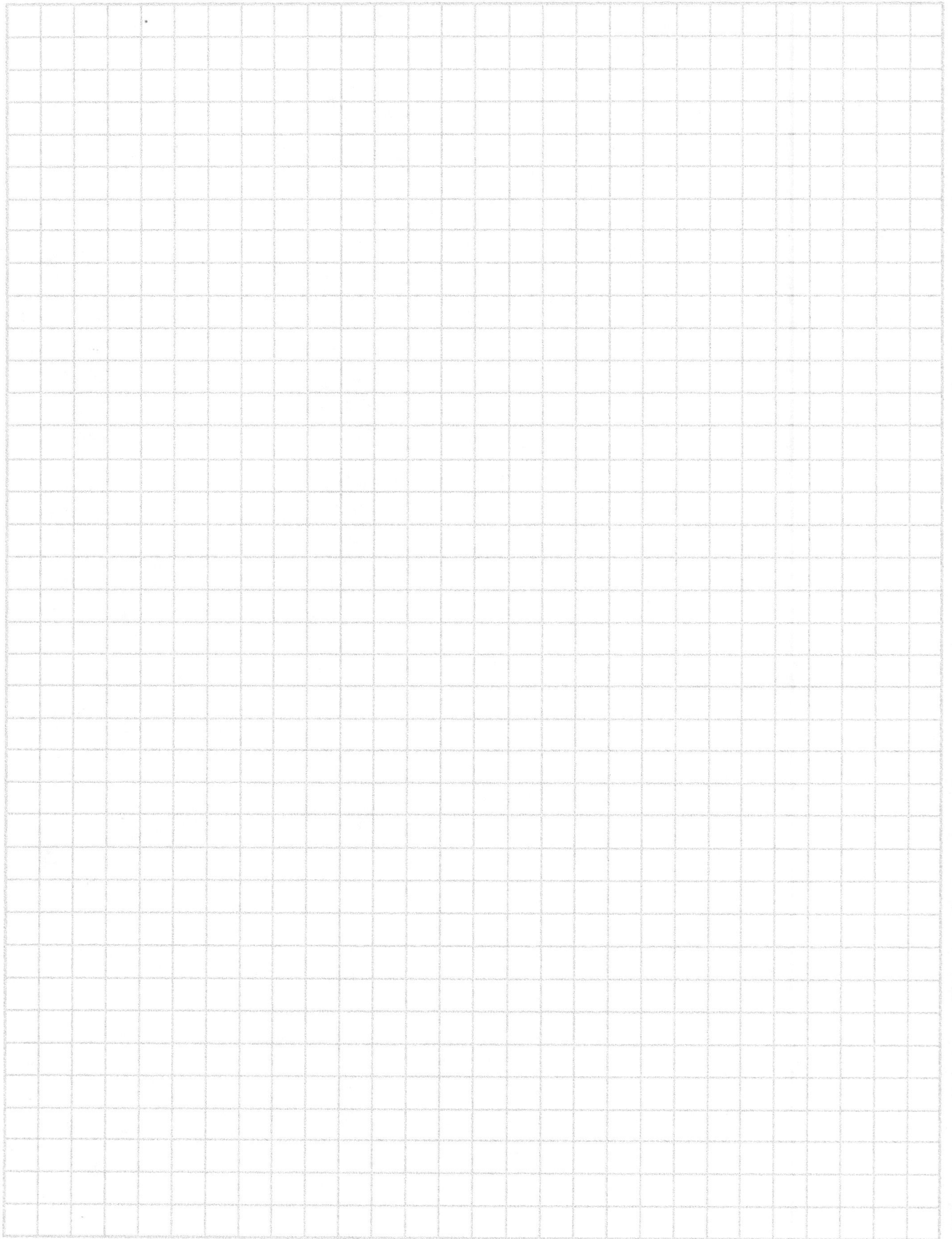

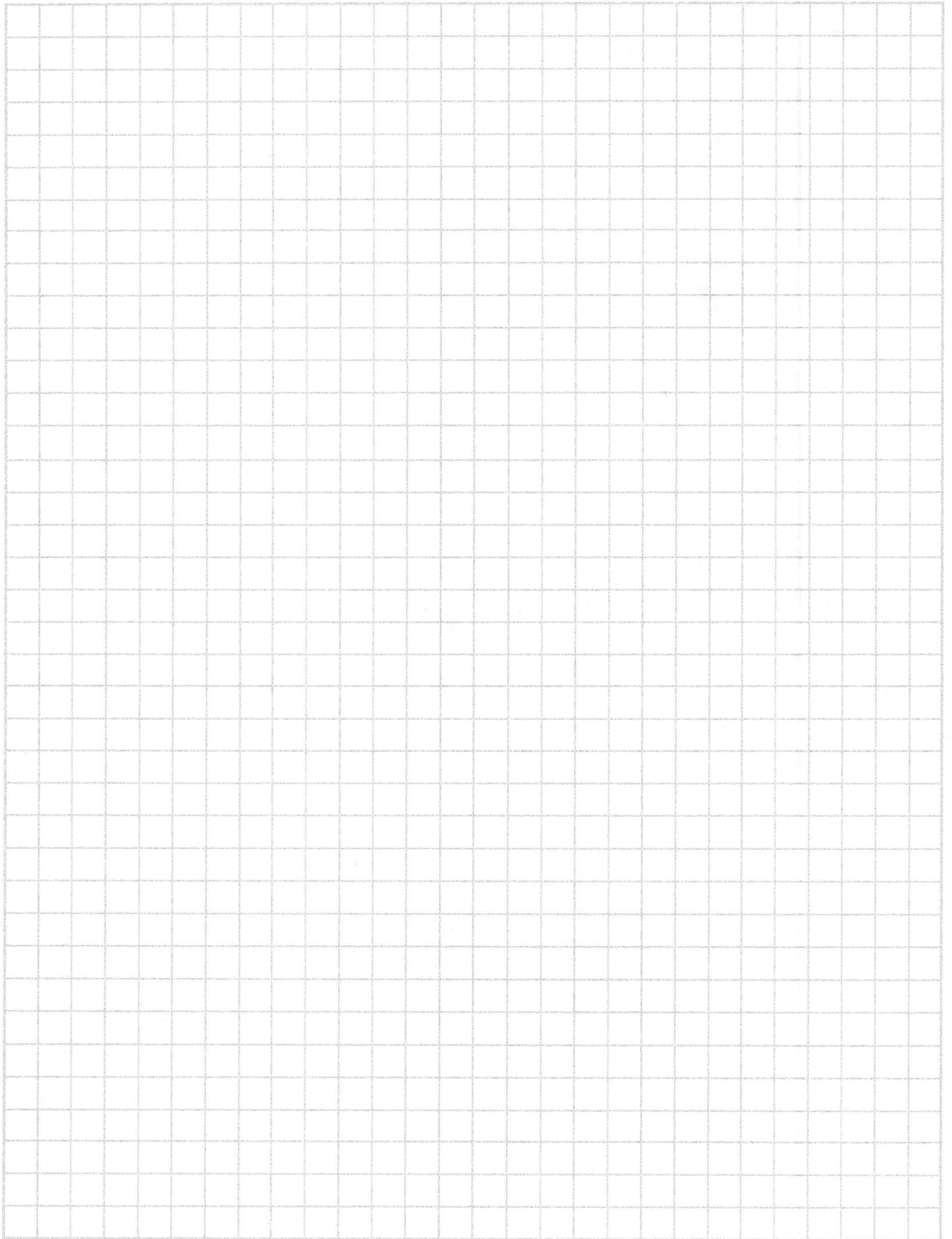

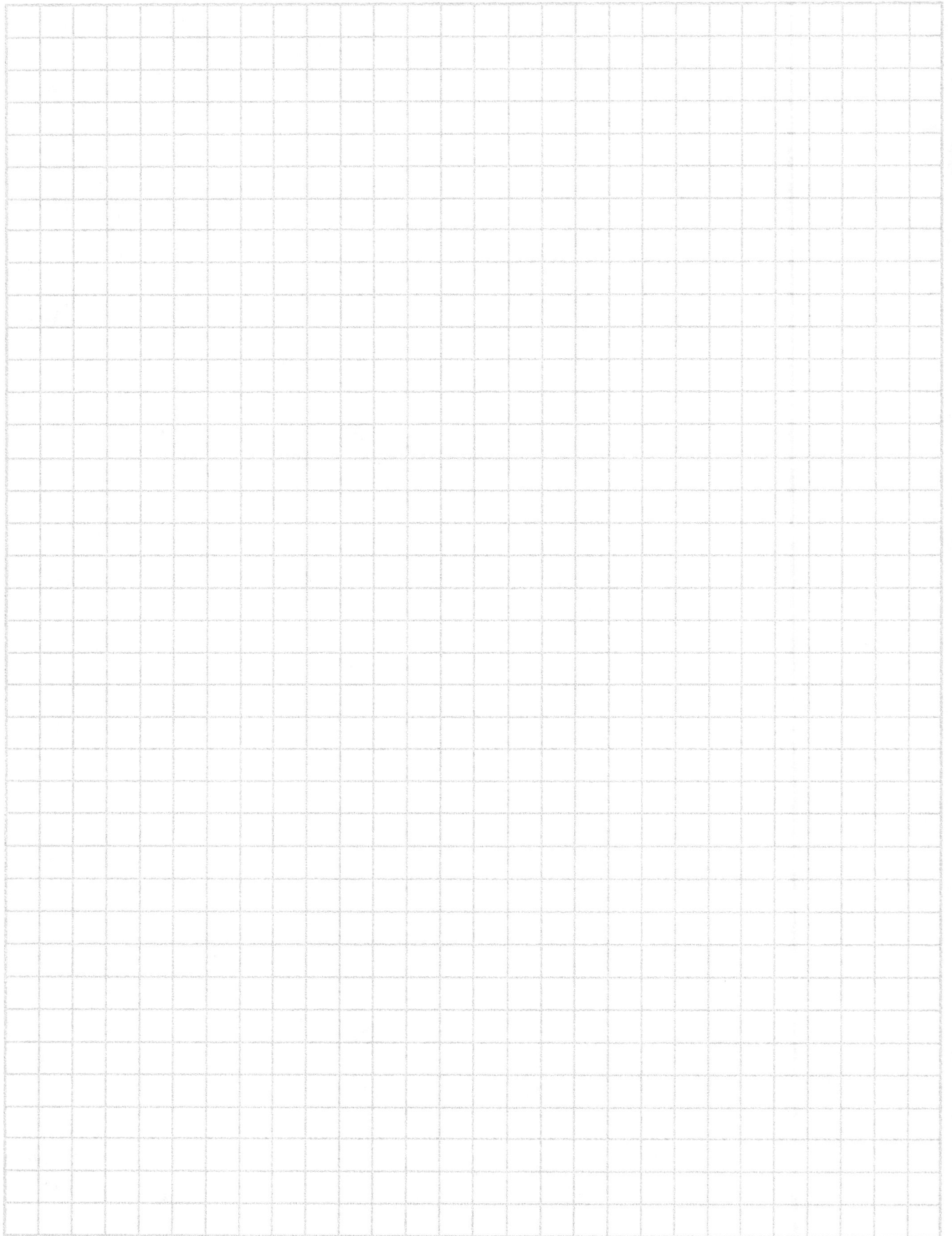

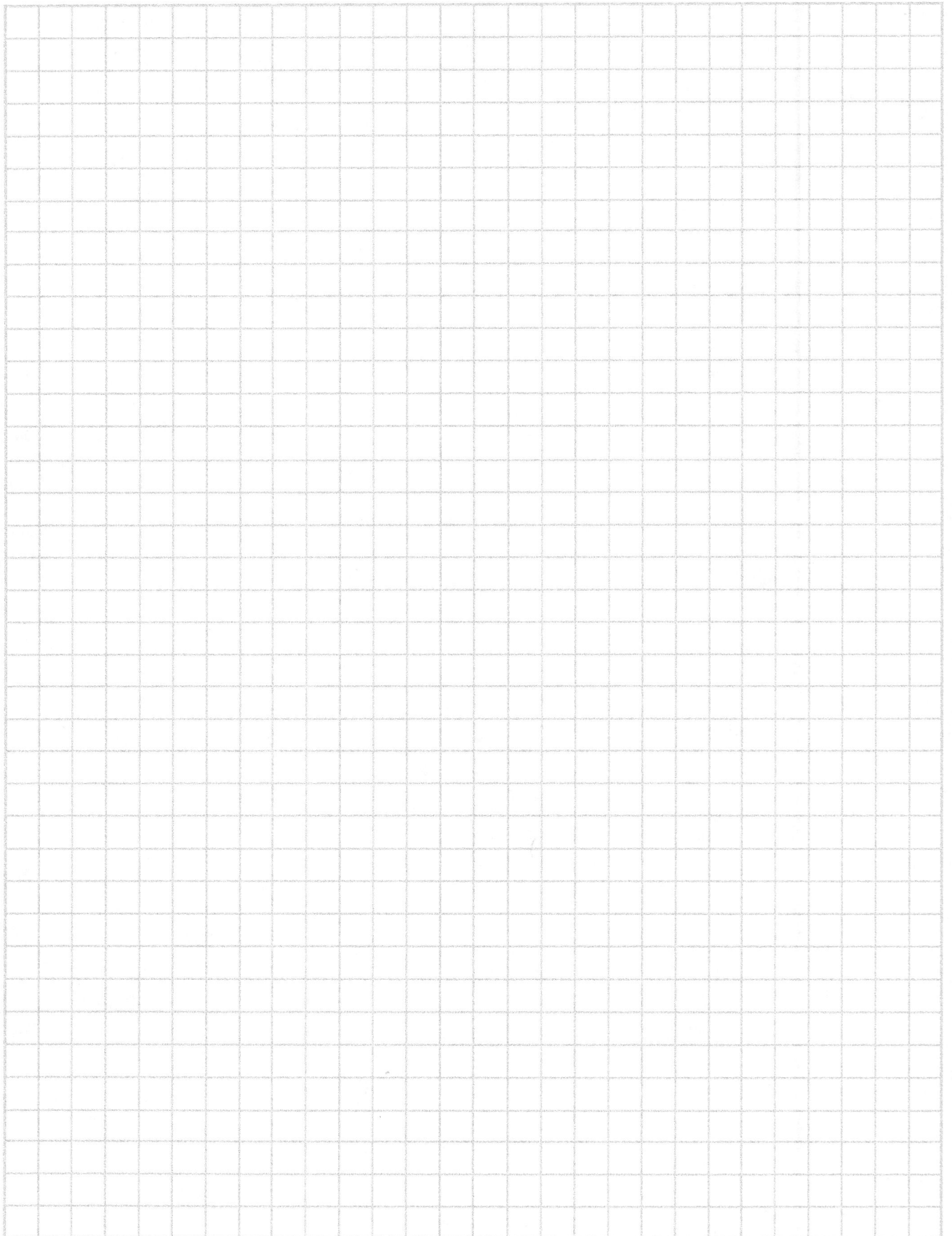

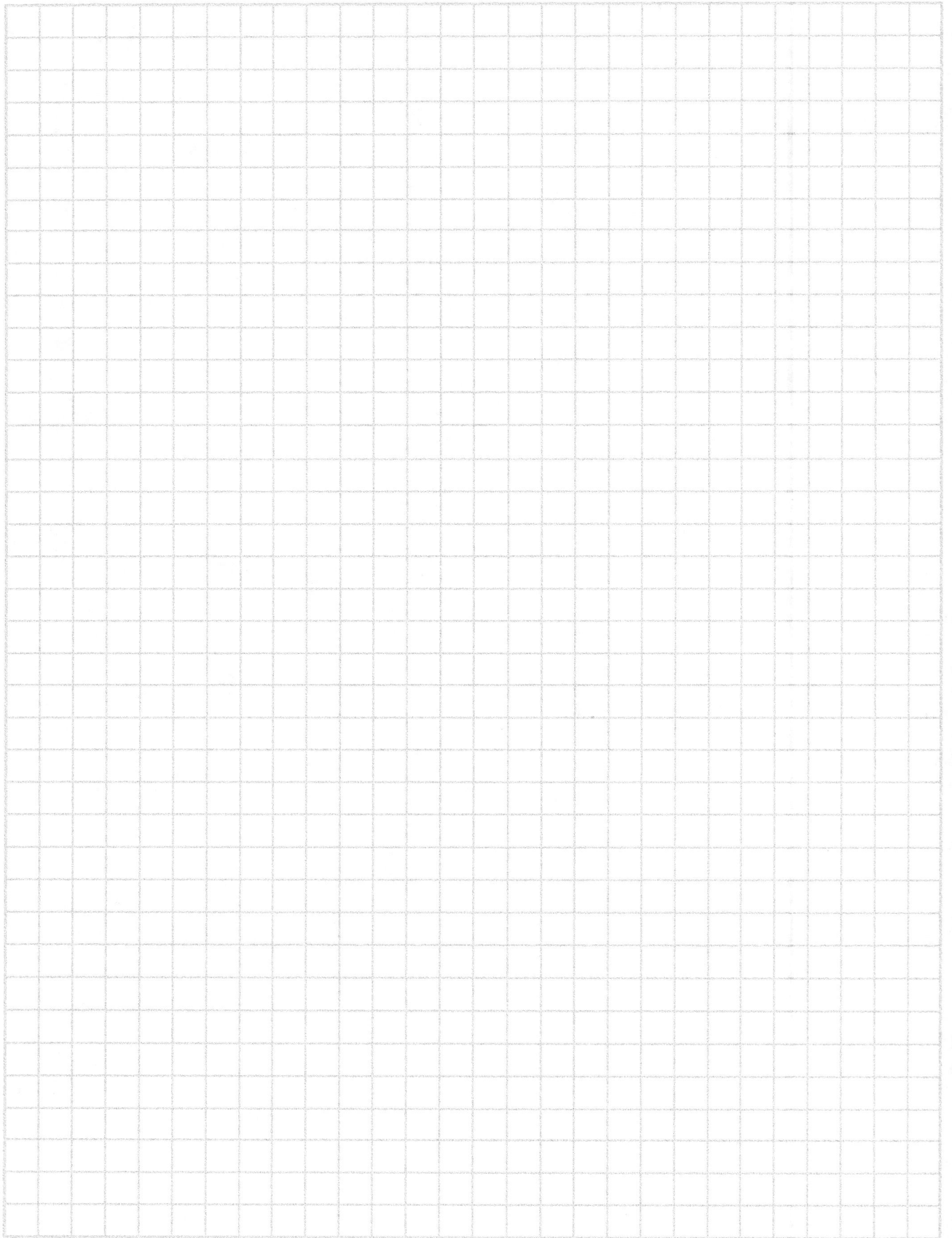

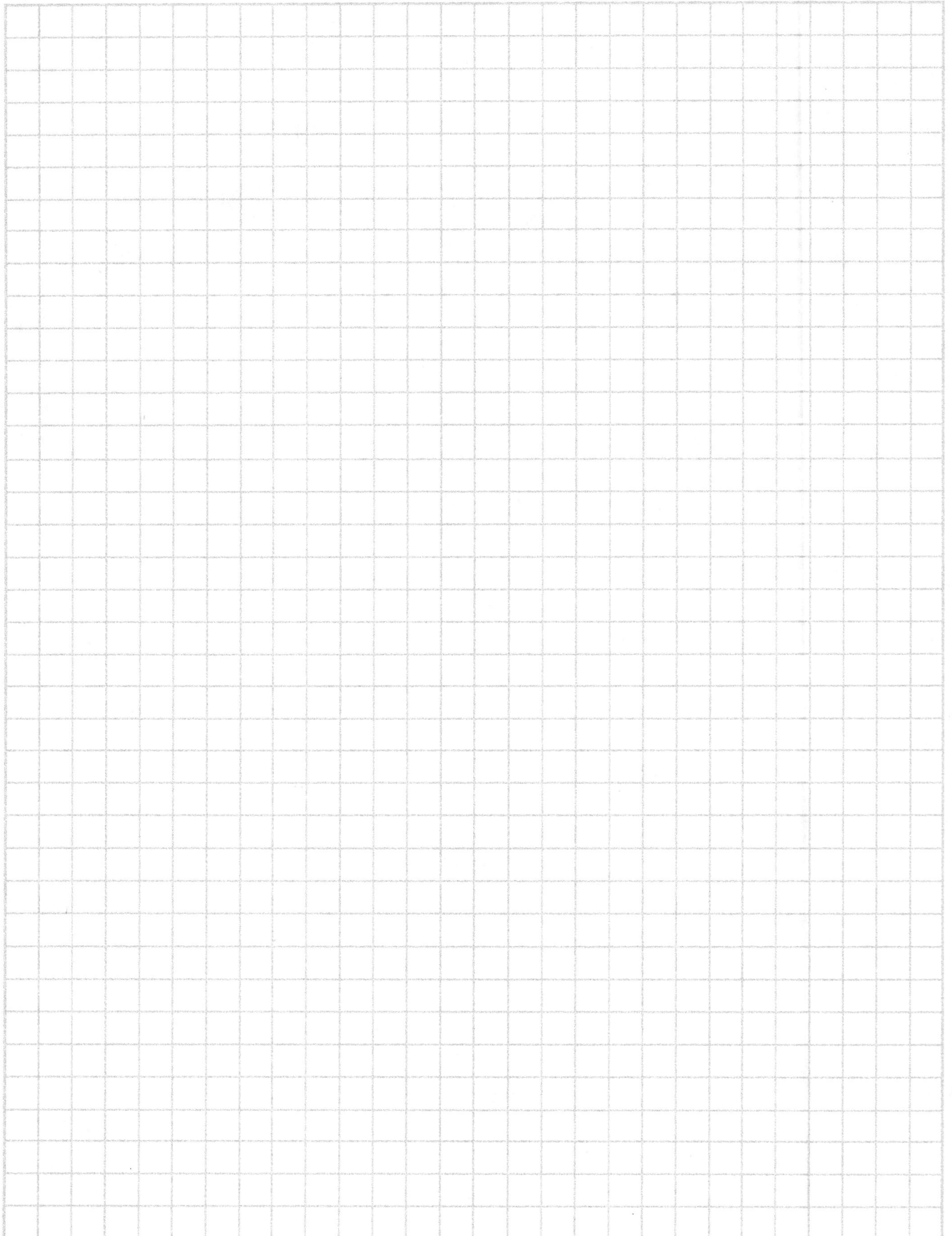

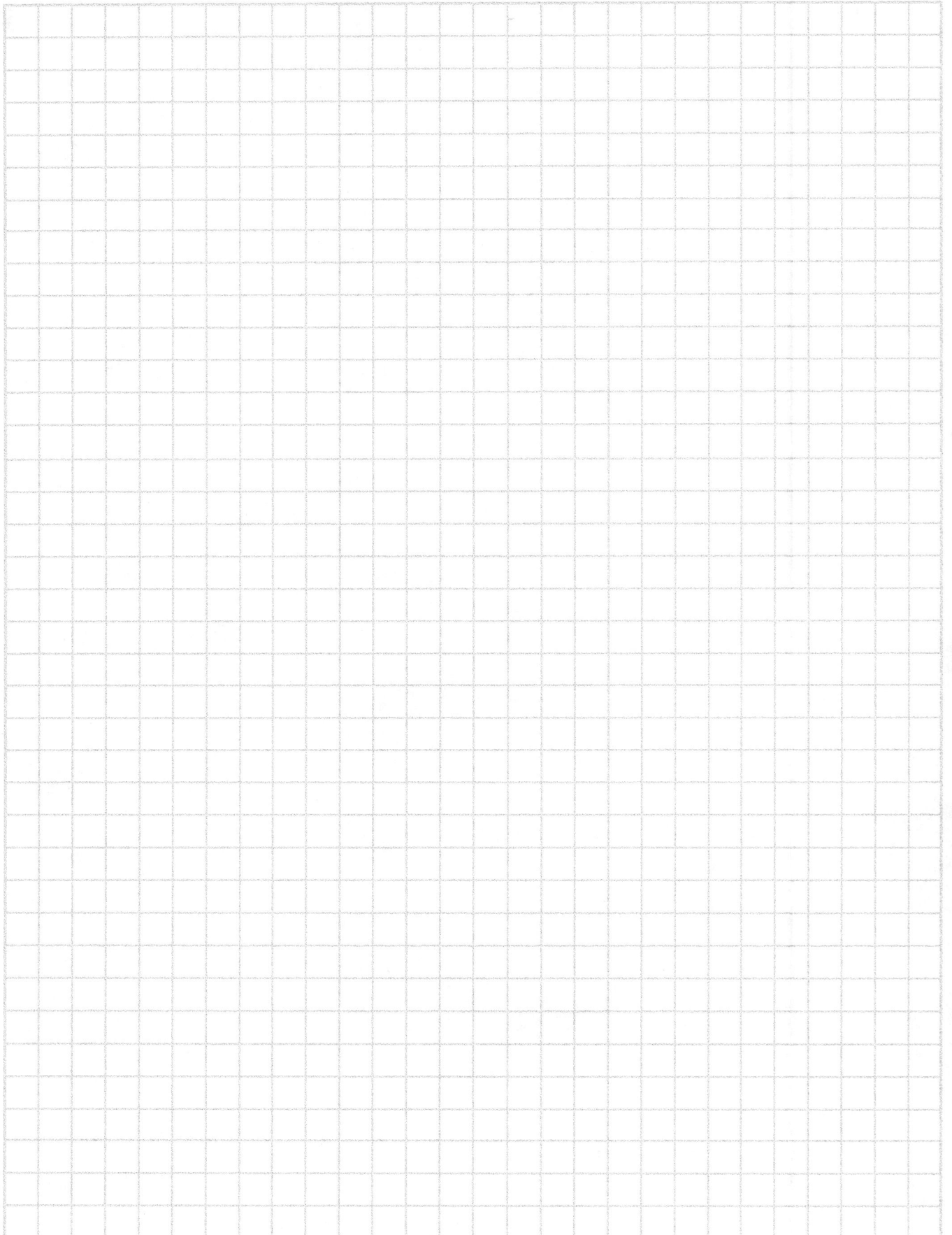

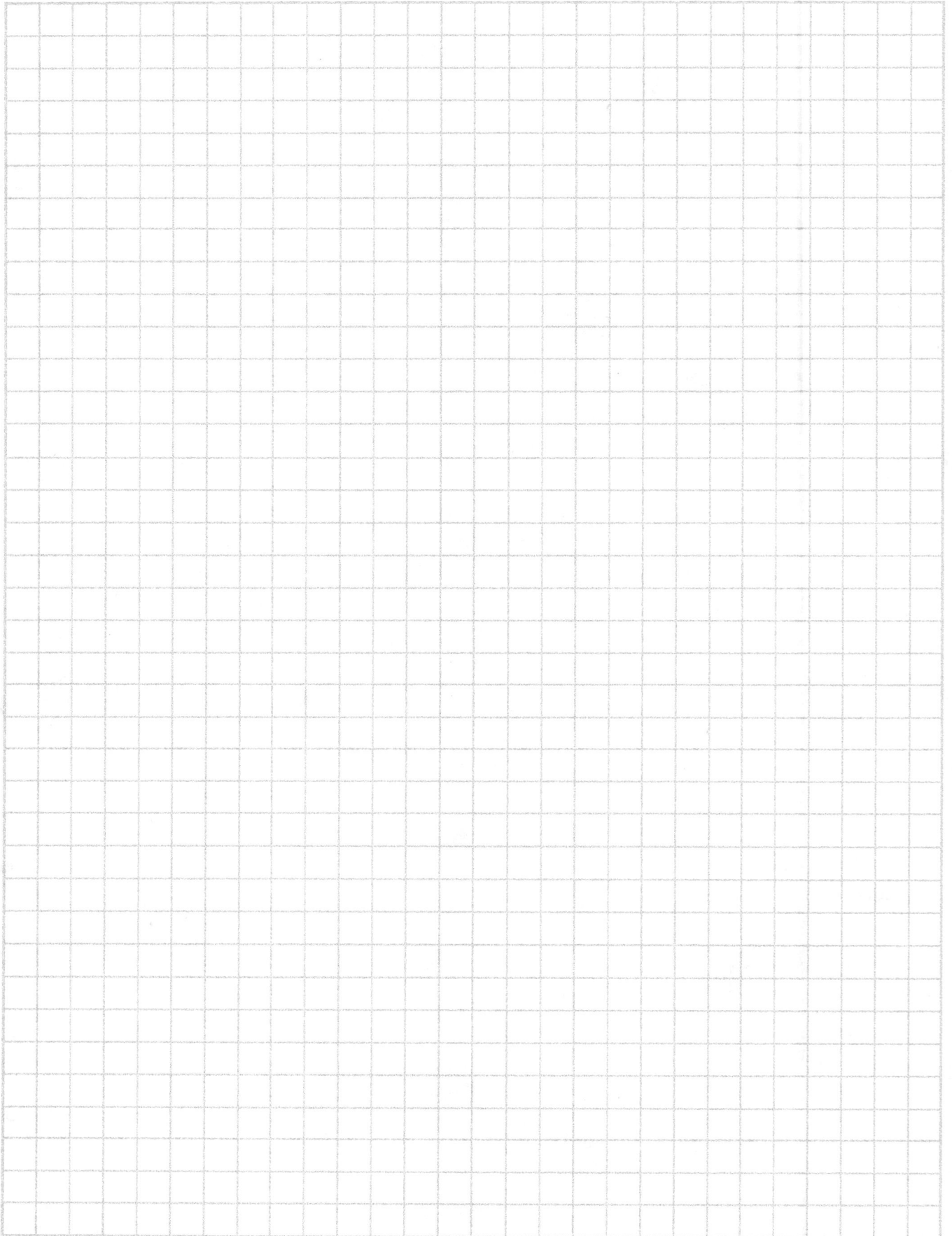

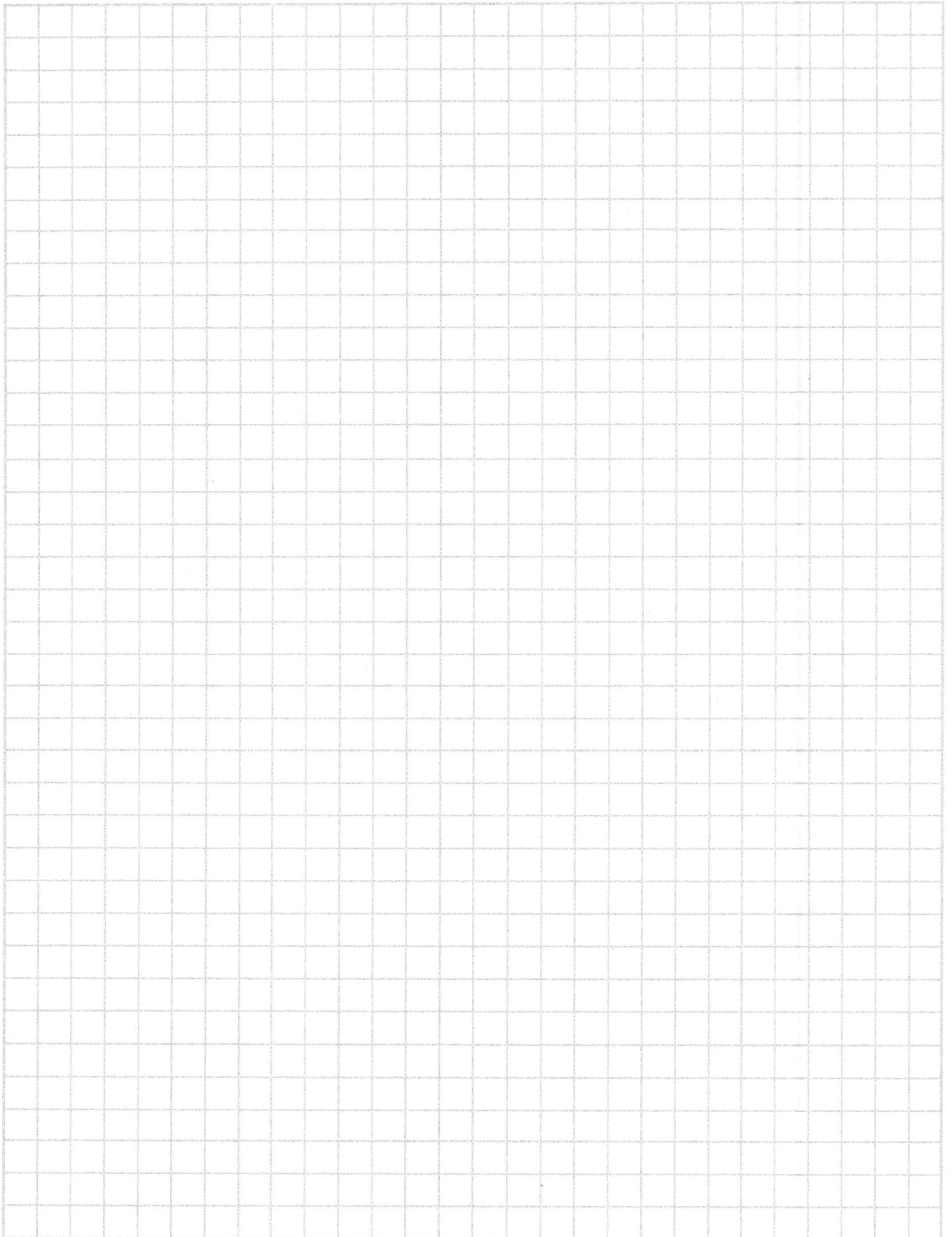

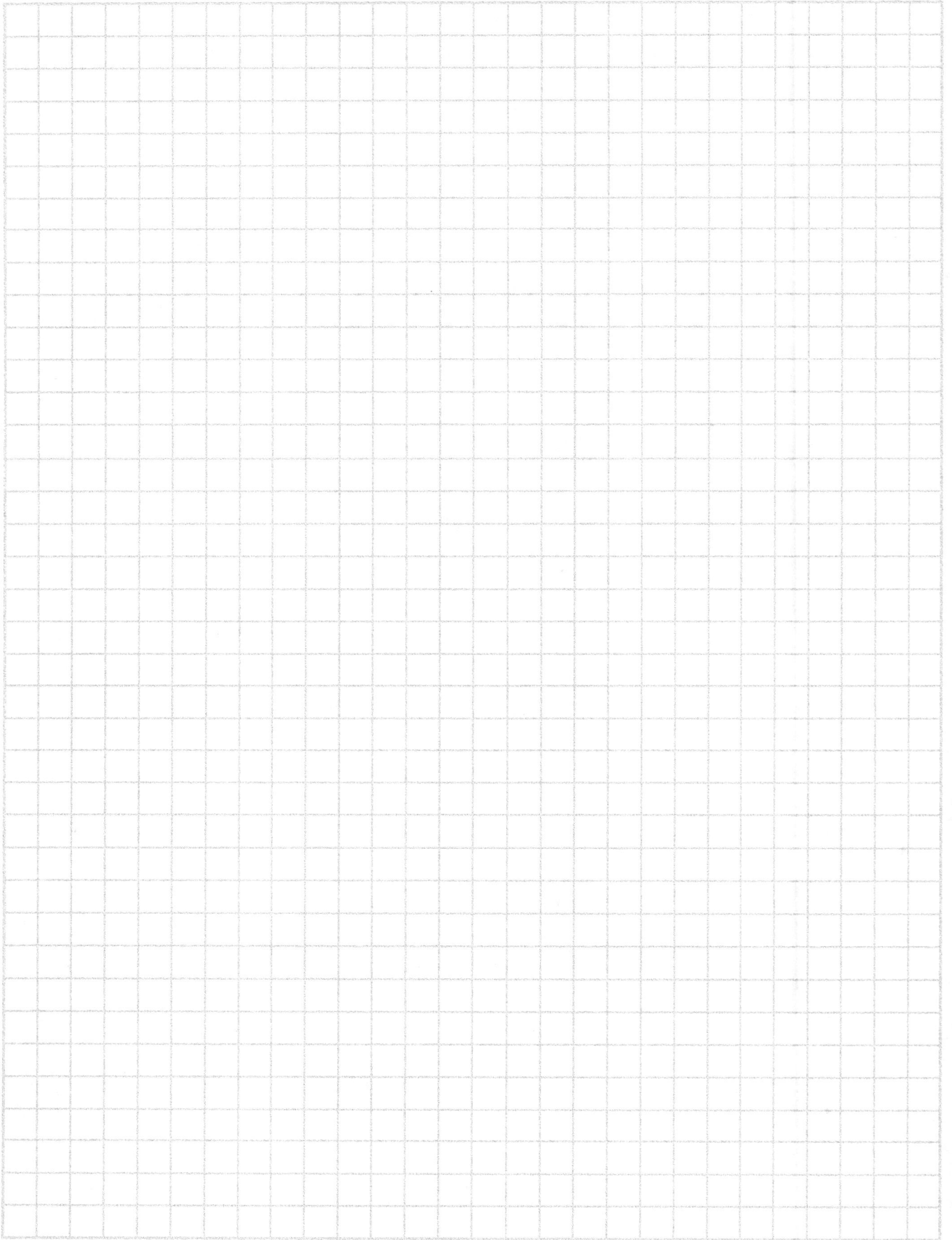

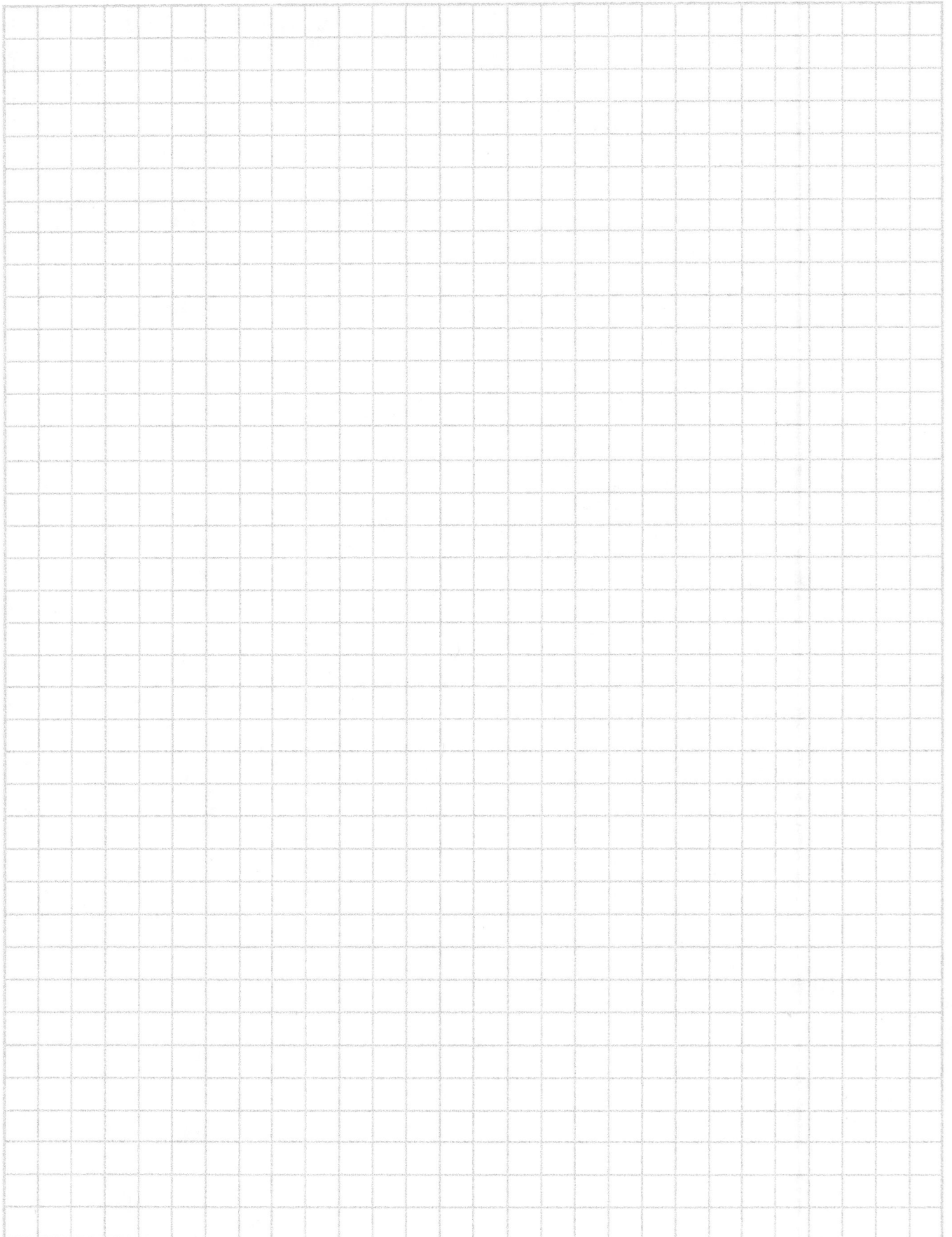

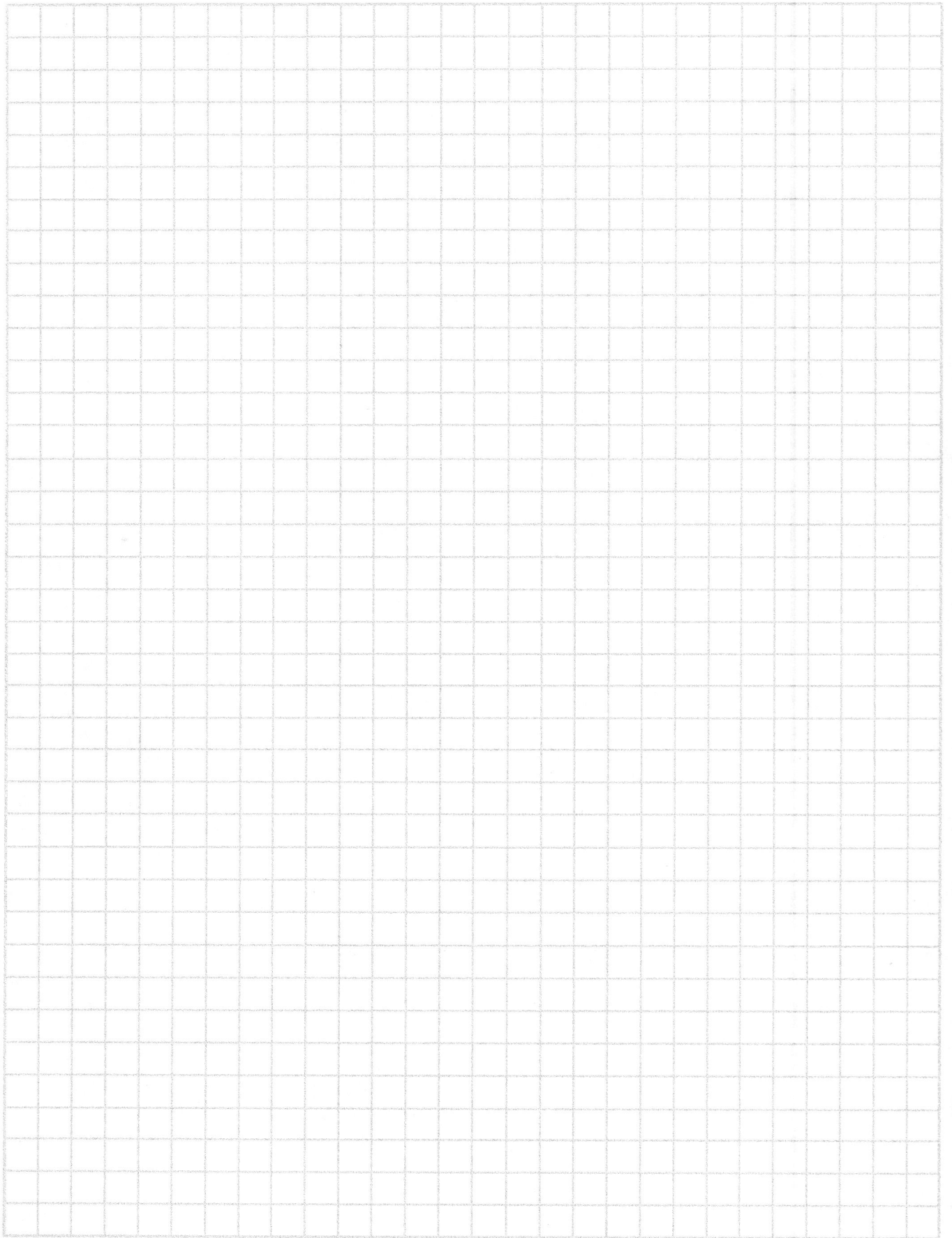

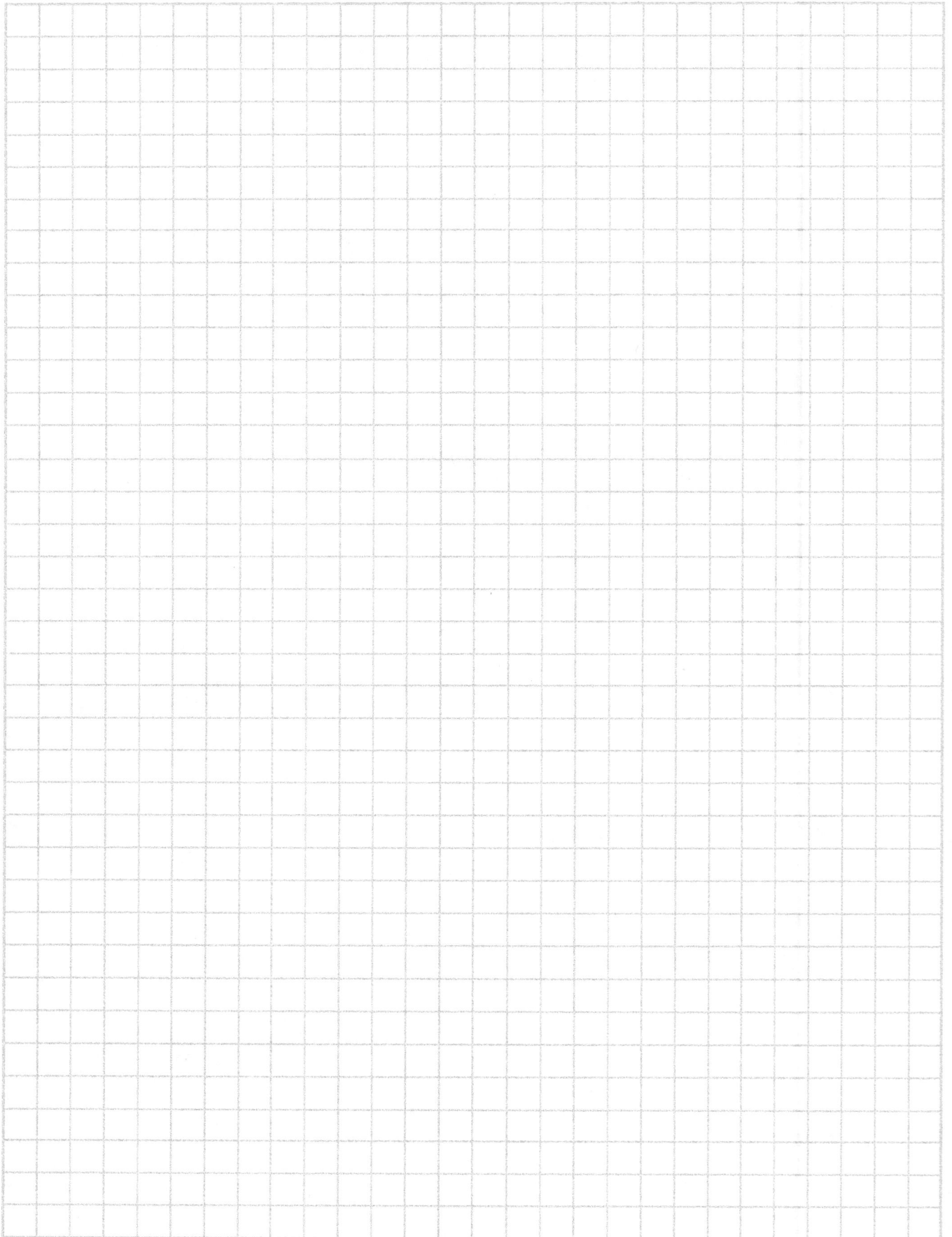

MATH ESSENTIALS
Math Resource Center

Important Geometry Formulas

Square

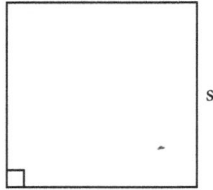

Perimeter: P = 4s
Area: $A = s^2$

Rectangle

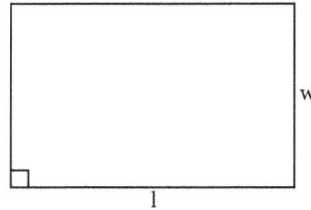

Perimeter: P = 2l + 2w
Area: A = lw

Triangle

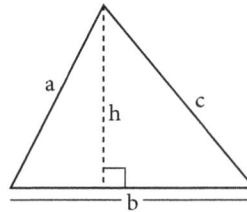

Perimeter: P = a + b + c
Area: $A = \frac{1}{2}bh$

Sum of Angles of Triangle

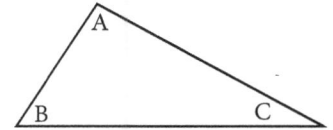

A + B + C = 180°
The sum of the measures of the three angles is 180°.

Right Triangle

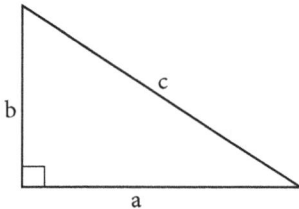

Perimeter: P = a + b + c
Area: $A = \frac{1}{2}ab$
One 90° (right) angle

Pythagorean Theorem (for right angles)

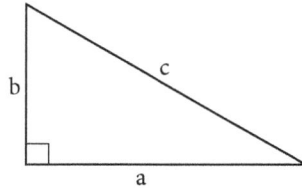

$a^2 + b^2 = c^2$

Isosceles Triangle

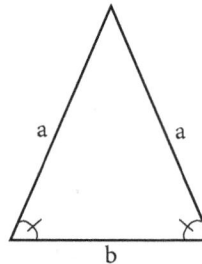

Triangle has two equal sides and two equal angles.

Equilateral Triangle

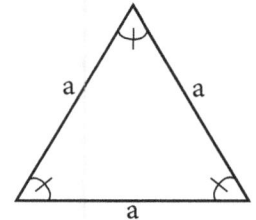

Triangle has three equal sides and three equal angles.

Trapezoid

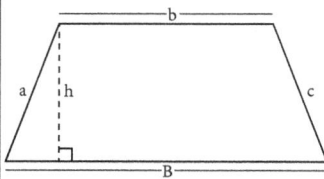

Perimeter: P = a + b + c + B
Area: $A = \frac{1}{2}h(B + b)$

Parallelogram

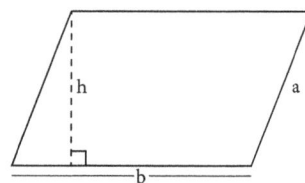

Perimeter: P = 2a + 2b
Area: A = bh

Circle

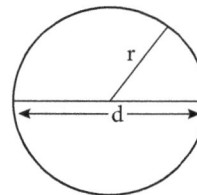

Circumference: C = π d
Area: $A = \pi r^2$

Rectangular Solid

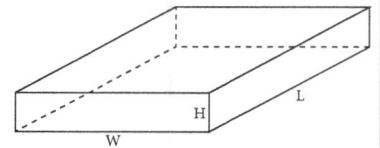

Volume: V = LWH
Surface Area:
S = 2LH + 2LW + 2WH

Cube

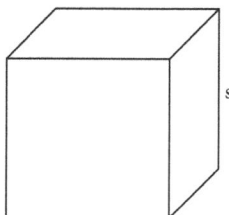

Volume: $V = s^3$

Cone

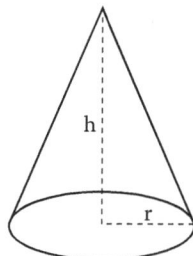

Volume: $V = \frac{1}{3}\pi r^2 h$

Right Circular Cylinder

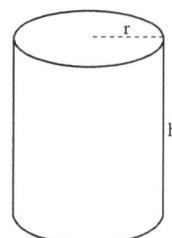

Volume: $V = \pi r^2 h$

Sphere

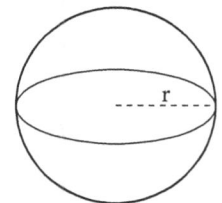

Volume: $V = \frac{4}{3}\pi r^3$

Glossary

absolute value   The distance of a number from 0 on the number line. The absolute value is always positive.

acute angle   An angle with a measure of less than 90 degrees.

adjacent   Next to.

algebraic expression   A mathematical expression that contains at least one variable.

angle   Any two rays that share an endpoint will form an angle.

associative properties   For any a, b, c:
addition: $(a + b) + c = a + (b + c)$
multiplication: $(ab)c = a(bc)$

B

base   The number being multiplied. In an exponential expression such as 4^2, 4 is the base.

C

coefficient   A number that multiplies the variable. In the term 7x, 7 is the coefficient of x.

commutative properties   For any a, b:
addition: $a + b = b + a$
multiplication: $ab = ba$

complementary angles   Two angles that have measures whose sum is 90 degrees.

congruent   Two figures having exactly the same size and shape.

coordinate plane   The plane which contains the x- and y-axes. It is divided into 4 quadrants. Also called coordinate system and coordinate grid.

coordinates   An ordered pair of numbers that identify a point on a coordinate plane.

D

data   Information that is organized for analysis.

degree   A unit that is used in measuring angles.

denominator   The bottom number of a fraction that tells the number of equal parts into which a whole is divided.

disjoint sets   Sets that have no members in common. {1,2,3} and {4,5,6} are disjoint sets.

Glossary

distributive property   For real numbers a, b, and c: a(b + c) = ab + ac.

element of a set   Member of a set.

empty set   The set that has no members. Also called the null set and written $\emptyset$ or { }.

equation   A mathematical sentence that contains an equal sign (=) and states that one expression is equal to another expression.

equivalent   Having the same value.

exponent   A number that indicates the number of times a given base is used as a factor. In the expression n^2, 2 is the exponent.

expression   Variables, numbers, and symbols that show a mathematical relationship.

extremes of a proportion   In the proportion $\frac{a}{b} = \frac{c}{d}$, a and d are the extremes.

factor   An integer that divides evenly into another.

finite   Something that is countable.

formula   A general mathematical statement or rule. Used often in algebra and geometry.

function   A set of ordered pairs that pairs each x-value with one and only one y-value.
        (0,2), (-1,6), (4,-2), (-3,4) is a function.

graph   To show points named by numbers or ordered pairs on a number line or coordinate plane.
        Also, a drawing to show the relationship between sets of data.

greatest common factor   The largest common factor of two or more numbers. Also written GCF.
        The greatest common factor of 15 and 25 is 5.

grouping symbols   Symbols that indicate the order in which mathematical operations should take place.
        Examples include parentheses ( ), brackets [ ], braces { }, and fraction bars —— .

hypotenuse   The side opposite the right angle in a right triangle.

I

identity properties of addition and multiplication   For any real number a:
addition: $a + 0 = 0 + a = a$
multiplication: $1 \times a = a \times 1 = a$

inequality   A mathematical sentence that states one expression is greater than or less than another.
Inequality symbols are read as follows: $<$ less than
$\leq$ less than or equal to
$>$ greater than
$\geq$ greater than or equal to

infinite   Having no boundaries or limits. Uncountable.

integers   Numbers in a set. {...-3, -2, -1, 0, 1, 2, 3...}

intersection of sets   If A and B are sets, then A intersection B is the set whose members are included in both sets A and B, and is written $A \cap B$. If set A = {1,2,3,4} and set B = {1,3,5}, then $A \cap B$ = {1,3}

inverse properties of addition and multiplication   For any number a:
addition: $a + -a = 0$
multiplication: $a \times 1/a = 1 \ (a \neq 0)$

inverse operations   Operations that "undo" each other. Addition and subtraction are inverse operations, and multiplication and division are inverse operations.

L

least common multiple   The least common multiple of two or more whole numbers is the smallest whole number, other than zero, that they all divide into evenly. Also written as LCM. The least common multiple of 12 and 8 is 24.

linear equation   An equation whose graph is a straight line.

M

mean   In statistics, the sum of a set of numbers divided by the number of elements in the set. Sometimes referred to as average.

means of a proportion   In the proportion $\frac{a}{b} = \frac{c}{d}$, b and c are the means.

median   In statistics, the middle number of a set of numbers when the numbers are arranged in order of least to greatest. If there are two middle numbers, find their mean.

mode   In statistics, the number that appears most frequently. Sometimes there is no mode. There may also be more than one mode.

multiple   The product of a whole number and another whole number.

Glossary

N

natural numbers Numbers in the set {1, 2, 3, 4, ...}. Also called counting numbers.

negative numbers Numbers that are less than zero.

null set The set that has no members. Also called the empty set and written Ø or { }.

number line A line that represents numbers as points.

numerator The top part of a fraction.

O

obtuse angle An angle whose measure is greater than 90° and less than 180°.

opposites Numbers that are the same distance from zero, but are on opposite sides of zero on a number line. 4 and -4 are opposites.

order of operations The order of steps to be used when simplifying expressions.
   1. Evaluate within grouping symbols.
   2. Eliminate all exponents.
   3. Multiply and divide in order from left to right.
   4. Add and subtract in order from left to right.

ordered pair A pair of numbers (x,y) that represent a point on the coordinate plane. The first number is the x-coordinate and the second number is the y-coordinate.

origin The point where the x-axis and the y-axis intersect in a coordinate plane. Written as (0,0).

outcome One of the possible events in a probability situation.

P

parallel lines Lines in a plane that do not intersect. They stay the same distance apart.

percent Hundredths or per hundred. Written %.

perimeter The distance around a figure.

perpendicular lines Lines in the same plane that intersect at a right (90°) angle.

pi The ratio of the circumference of a circle to its diameter. Written π. The approximate value for π is 3.14 as a decimal and $\frac{22}{7}$ as a fraction.

plane A flat surface that extends infinitely in all directions.

point An exact position in space. Points also represent numbers on a number line or coordinate plane.

positive number Any number that is greater than 0.

power An exponent.

prime number A whole number greater than 1 whose only factors are 1 and itself.

probability What chance, or how likely it is for an event to occur. It is the ratio of the ways a certain outcome can occur and the number of possible outcomes.

proportion An equation that states that two ratios are equal. $\frac{4}{8} = \frac{2}{4}$ is a proportion.

Pythagorean theorem In a right triangle, if c is the hypotenuse, and a and b are the other two legs, then $a^2 + b^2 = c^2$.

Q

quadrant One of the four regions into which the x-axis and y-axis divide a coordinate plane.

$$\begin{array}{c|c} \text{II} & \text{I} \\ \hline \text{III} & \text{IV} \end{array}$$

R

range The difference between the greatest number and the least number in a set of numbers.

ratio A comparison of two numbers using division. Written a:b, a to b, and a/b.

reciprocals Two numbers whose product is 1. $\frac{2}{3}$ and $\frac{3}{2}$ are reciprocals because $\frac{2}{3} \times \frac{3}{2} = 1$.

reduce To express a fraction in its lowest terms.

relation Any set of ordered pairs.

right angle An angle that has a measure of 90°.

rise The change in y going from one point to another on a coordinate plane. The vertical change.

run The change in x going from one point to another on a coordinate plane. The horizontal change.

S

scientific notation A number written as the product of a numbers between 1 and 10 and a power of ten. In scientific notation, $7,000 = 7 \times 10^3$.

set A well-defined collection of objects.

slope Refers to the slant of a line. It is the ratio of rise to run.

Glossary

solution A number that can be substituted for a variable to make an equation true.

square root Written $\sqrt{}$. The $\sqrt{36}$ = 6 because 6 × 6 = 36.

statistics Involves data that is gathered about people or things and is used for analysis.

subset If all the members of set A are members of set B, then set A is a subset of set B. Written A⊂B. If set A = {1,2,3} and set B = {0,1,2,3,5,8}, set A is a subset of set B because all of the members of a set A are also members of set B.

U

union of sets If A and B are sets, the union of set A and set B is the set whose members are included in set A, or set B, or both set A and set B. A union B is written A∪B. If set = {1,2,3,4} and set B = {1,3,5,7}, then A∪B = {1,2,3,4,5,7}.

universal set The set which contains all the other sets which are under consideration.

V

variable A letter that represents a number.

Venn diagram A type of diagram that shows how certain sets are related.

vertex The point at which two lines, line segments, or rays meet to form an angle.

W

whole number Any number in the set {0, 1, 2, 3, 4...}

X

x-axis The horizontal axis on a coordinate plane.

x-coordinate The first number in an ordered pair. Also called the abscissa.

Y

y-axis The vertical axis on a coordinate plane.

y-coordinate The second number in an ordered pair. Also called the ordinate.

Important Symbols

$<$	less than	π	pi
$\leq$	less than or equal to	{ }	set
$>$	greater than	| |	absolute value
$\geq$	greater than or equal to	$.\overline{n}$	repeating decimal symbol
$=$	equal to	$1/a$	the reciprocal of a number
$\neq$	not equal to	%	percent
$\cong$	congruent to	(x,y)	ordered pair
( )	parenthesis	$\perp$	perpendicular
[ ]	brackets	| |	parallel to
{ }	braces	$\angle$	angle
...	and so on	$\in$	element of
• or ×	multiply	$\notin$	not an element of
∞	infinity	$\cap$	intersection
a^n	the nth power of a number	$\cup$	union
$\sqrt{}$	square root	$\subset$	subset of
Ø, { }	the empty set or null set	$\not\subset$	not a subset of
$\therefore$	therefore	$\triangle$	triangle
°	degree		

Multiplication Table

x	2	3	4	5	6	7	8	9	10	11	12
2	4	6	8	10	12	14	16	18	20	22	24
3	6	9	12	15	18	21	24	27	30	33	36
4	8	12	16	20	24	28	32	36	40	44	48
5	10	15	20	25	30	35	40	45	50	55	60
6	12	18	24	30	36	42	48	54	60	66	72
7	14	21	28	35	42	49	56	63	70	77	84
8	16	24	32	40	48	56	64	72	80	88	96
9	18	27	36	45	54	63	72	81	90	99	108
10	20	30	40	50	60	70	80	90	100	110	120
11	22	33	44	55	66	77	88	99	110	121	132
12	24	36	48	60	72	84	96	108	120	132	144

Commonly Used Prime Numbers

2	3	5	7	11	13	17	19	23	29
31	37	41	43	47	53	59	61	67	71
73	79	83	89	97	101	103	107	109	113
127	131	137	139	149	151	157	163	167	173
179	181	191	193	197	199	211	223	227	229
233	239	241	251	257	263	269	271	277	281
283	293	307	311	313	317	331	337	347	349
353	359	367	373	379	383	389	397	401	409
419	421	431	433	439	443	449	457	461	463
467	479	487	491	499	503	509	521	523	541
547	557	563	569	571	577	587	593	599	601
607	613	617	619	631	641	643	647	653	659
661	673	677	683	691	701	709	719	727	733
739	743	751	757	761	769	773	787	797	809
811	821	823	827	829	839	853	857	859	863
877	881	883	887	907	911	919	929	937	941
947	953	967	971	977	983	991	997	1009	1013

Squares and Square Roots

No.	Square	Square Root	No.	Square	Square Root	No.	Square	Square Root
1	1	1.000	51	2,601	7.141	101	10201	10.050
2	4	1.414	52	2,704	7.211	102	10,404	10.100
3	9	1.732	53	2,809	7.280	103	10,609	10.149
4	16	2.000	54	2,916	7.348	104	10,816	10.198
5	25	2.236	55	3,025	7.416	105	11,025	10.247
6	36	2.449	56	3,136	7.483	106	11,236	10.296
7	49	2.646	57	3,249	7.550	107	11,449	10.344
8	64	2.828	58	3,364	7.616	108	11,664	10.392
9	81	3.000	59	3,481	7.681	109	11,881	10.440
10	100	3.162	60	3,600	7.746	110	12,100	10.488
11	121	3.317	61	3,721	7.810	111	12,321	10.536
12	144	3.464	62	3,844	7.874	112	12,544	10.583
13	169	3.606	63	3,969	7.937	113	12,769	10.630
14	196	3.742	64	4,096	8.000	114	12,996	10.677
15	225	3.873	65	4,225	8.062	115	13,225	10.724
16	256	4.000	66	4,356	8.124	116	13,456	10.770
17	289	4.123	67	4,489	8.185	117	13,689	10.817
18	324	4.243	68	4,624	8.246	118	13,924	10.863
19	361	4.359	69	4,761	8.307	119	14,161	10.909
20	400	4.472	70	4,900	8.367	120	14,400	10.954
21	441	4.583	71	5,041	8.426	121	14,641	11.000
22	484	4.690	72	5,184	8.485	122	14,884	11.045
23	529	4.796	73	5,329	8.544	123	15,129	11.091
24	576	4.899	74	5,476	8.602	124	15,376	11.136
25	625	5.000	75	5,625	8.660	125	15,625	11.180
26	676	5.099	76	5,776	8.718	126	15,876	11.225
27	729	5.196	77	5,929	8.775	127	16,129	11.269
28	784	5.292	78	6,084	8.832	128	16,384	11.314
29	841	5.385	79	6,241	8.888	129	16,641	11.358
30	900	5.477	80	6,400	8.944	130	16,900	11.402
31	961	5.568	81	6,561	9.000	131	17,161	11.446
32	1,024	5.657	82	6,724	9.055	132	17,424	11.489
33	1,089	5.745	83	6,889	9.110	133	17,689	11.533
34	1,156	5.831	84	7,056	9.165	134	17,956	11.576
35	1,225	5.916	85	7,225	9.220	135	18,225	11.619
36	1,296	6.000	86	7,396	9.274	136	18,496	11.662
37	1,369	6.083	87	7,569	9.327	137	18,769	11.705
38	1,444	6.164	88	7,744	9.381	138	19,044	11.747
39	1,521	6.245	89	7,921	9.434	139	19,321	11.790
40	1,600	6.325	90	8,100	9.487	140	19,600	11.832
41	1,681	6.403	91	8,281	9.539	141	19,881	11.874
42	1,764	6.481	92	8,464	9.592	142	20,164	11.916
43	1,849	6.557	93	8,649	9.644	143	20,449	11.958
44	1,936	6.633	94	8,836	9.695	144	20,736	12.000
45	2,025	6.708	95	9,025	9.747	145	21,025	12.042
46	2,116	6.782	96	9,216	9.798	146	21,316	12.083
47	2,209	6.856	97	9,409	9.849	147	21,609	12.124
48	2,304	6.928	98	9,604	9.899	148	21,904	12.166
49	2,401	7.000	99	9,801	9.950	149	22,201	12.207
50	2,500	7.071	100	10,000	10.000	150	22,500	12.247

Fraction/Decimal Equivalents

Fraction	Decimal	Fraction	Decimal
$\frac{1}{2}$	0.5	$\frac{5}{10}$	0.5
$\frac{1}{3}$	$0.\overline{3}$	$\frac{6}{10}$	0.6
$\frac{2}{3}$	$0.\overline{6}$	$\frac{7}{10}$	0.7
$\frac{1}{4}$	0.25	$\frac{8}{10}$	0.8
$\frac{2}{4}$	0.5	$\frac{9}{10}$	0.9
$\frac{3}{4}$	0.75	$\frac{1}{16}$	0.0625
$\frac{1}{5}$	0.2	$\frac{2}{16}$	0.125
$\frac{2}{5}$	0.4	$\frac{3}{16}$	0.1875
$\frac{3}{5}$	0.6	$\frac{4}{16}$	0.25
$\frac{4}{5}$	0.8	$\frac{5}{16}$	0.3125
$\frac{1}{8}$	0.125	$\frac{6}{16}$	0.375
$\frac{2}{8}$	0.25	$\frac{7}{16}$	0.4375
$\frac{3}{8}$	0.375	$\frac{8}{16}$	0.5
$\frac{4}{8}$	0.5	$\frac{9}{16}$	0.5625
$\frac{5}{8}$	0.625	$\frac{10}{16}$	0.625
$\frac{6}{8}$	0.75	$\frac{11}{16}$	0.6875
$\frac{7}{8}$	0.875	$\frac{12}{16}$	0.75
$\frac{1}{10}$	0.1	$\frac{13}{16}$	0.8125
$\frac{2}{10}$	0.2	$\frac{14}{16}$	0.875
$\frac{3}{10}$	0.3	$\frac{15}{16}$	0.9375
$\frac{4}{10}$	0.4